Signals and Communication Technology

This series is devoted to fundamentals and applications of modern methods of signal processing and cutting-edge communication technologies. The main topics are information and signal theory, acoustical signal processing, image processing and multimedia systems, mobile and wireless communications, and computer and communication networks. Volumes in the series address researchers in academia and industrial R&D departments. The series is application-oriented. The level of presentation of each individual volume, however, depends on the subject and can range from practical to scientific.

"Signals and Communication Technology" is indexed by Scopus.

More information about this series at http://www.springer.com/series/4748

Cheng Chi

Underwater Real-Time 3D Acoustical Imaging

Theory, Algorithm and System Design

Springer

Cheng Chi
Acoustic Research Laboratory, Tropical
Marine Science Institute
National University of Singapore
Singapore

ISSN 1860-4862 ISSN 1860-4870 (electronic)
Signals and Communication Technology
ISBN 978-981-13-3743-7 ISBN 978-981-13-3744-4 (eBook)
https://doi.org/10.1007/978-981-13-3744-4

Library of Congress Control Number: 2018968372

This Springer imprint is published by the registered company Springer Nature Singapore Pte Ltd.
The registered company address is: 152 Beach Road, #21-01/04 Gateway East, Singapore 189721, Singapore

To my beloved Yu, my parents and sister

Preface

Underwater real-time three-dimensional (3-D) acoustical imaging systems are able to capture real-time 3-D acoustical video. Other underwater acoustical imaging systems such as side-scan and multi-beam sonars can only obtain two-dimensional (2-D) images. Additionally, compared to underwater optical cameras, underwater real-time 3-D acoustical imaging systems achieve much longer imaging distance. Thus, underwater real-time 3-D acoustical imaging systems are becoming increasingly important in many applications such as underwater construction, pipe inspection, dredging, archaeology, anti-terrorist and diver detection.

This book will cover the theory, algorithms and system design of underwater real-time 3-D acoustical imaging. The existing underwater real-time 3-D acoustical imaging systems are narrowband. The techniques involved in developing narrowband systems, such design of large sparse 2-D arrays, fast beamforming are presented in our book. More importantly, this book summarizes the recent advances in wideband and ultrawideband underwater real-time 3-D acoustical imaging, which will be very useful for developing next-generation underwater real-time 3-D acoustical imaging systems. The simulation technique of underwater real-time 3-D acoustical imaging is also given in this book. It will help readers to learn and develop underwater real-time 3-D acoustical imaging systems fast.

In Chap. 1, this book presents an overview of underwater real-time 3-D acoustical imaging. The basic theories of underwater real-time 3-D acoustical imaging are introduced in Chap. 2. Chapter 3 shows the fast 3-D beamforming methods for underwater real-time 3-D acoustical imaging. The design techniques of large sparse 2-D arrays of underwater 3-D acoustical imaging, including narrowband, wideband and ultrawideband, are presented and discussed in Chap. 4. The simulation technique for designing these 3-D systems is given in Chap. 5. The steps of designing underwater real-time 3-D acoustical systems are presented in Chap. 6. Finally, this book outlines the future research potentials in Chap. 7.

Singapore
November 2018

Cheng Chi

Acknowledgements

The author would like to express my sincere thanks to Prof. Zhaohui Li, Prof. Renqian Wang, Prof. Qihu Li, Prof. Jiyuan Liu, Dr. Peng Wang, Dr. Jian Cui, Dr. Yang Zhang, Dr. Yu Hao and Dr. Pallayil Venugopalan.

The author thanks the editors of this series and the Springer team for their valuable guidance and assistance.

Contents

Chapter 1
Introduction

Abstract This chapter introduces the existing underwater real-time three-dimensional (3-D) acoustical imaging systems, and summarizes key techniques such as two-dimensional (2-D) array design and fast beamforming in developing the systems. This chapter also points out that all the practical developed 3-D systems work in narrowband, and next-generation underwater real-time 3-D imaging system should be wideband. The structure of this book is also shown.

Keywords Acoustical imaging · Fast beamforming · Real-time processing · 2-D array design · 3-D imaging system

1.1 Underwater Real-Time 3-D Acoustical Imaging Systems

With the growing demand for exploitation of subsea resources, underwater investigation is becoming more and more important. An underwater real-time 3-D acoustical imaging system can generate a 3-D oceanic environment image beyond the optical visibility range in a very short time [1–4]. Thus, underwater real-time 3-D acoustical imaging systems play an important role in underwater investigation. Real-time imaging means that the systems have to generate a 3-D image in real time. To keep consistent with 'optical' imaging systems, the word 'acoustical' is used to describe underwater 3-D acoustical imaging systems. Except this, 'acoustic' is used more frequently in this book.

According to [1, 2], underwater real-time 3-D acoustical imaging systems fall into three categories: (1) acoustical lens; (2) acoustical holography; (3) digital 3-D beamforming. Currently, the priority should be given to digital 3-D beamforming. In the last two decades, most of the developed underwater real-time 3-D imaging systems employ digital 3-D beamforming. The reasons why acoustical lens and holography are not used will be discussed in Chap. 2.

Figure 1.1 shows the block diagram of underwater real-time 3-D acoustical imaging systems. The wet ends of Coda Echoscope 3-D real-time system [5], and the system in [2] are given in Fig. 1.2. From Fig. 1.2, it can be seen that the transmit-

C. Chi, *Underwater Real-Time 3D Acoustical Imaging*, Signals and Communication Technology, https://doi.org/10.1007/978-981-13-3744-4_1

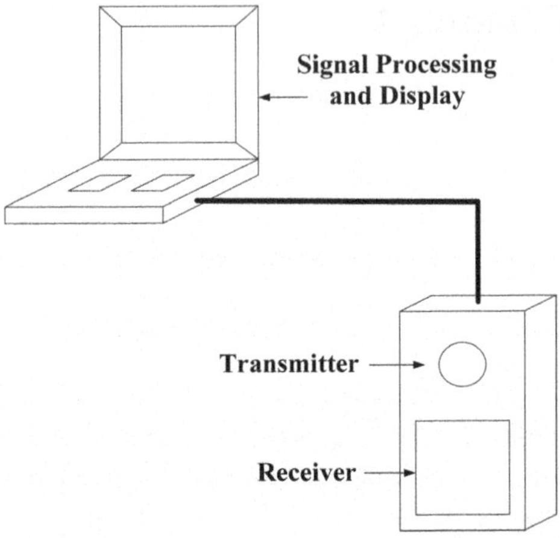

Fig. 1.1 Block diagram of an underwater real-time 3-D acoustical imaging system

ter is an omnidirectional projector. The reason the projectors are omnidirectional is explained as follows. The maximum range of underwater 3-D acoustical imaging systems exceeds several tens of meters generally [1–5]. If the transmitter is directional, we need to use the directional transmitter to scan a whole imaging 3-D scene mechanically or digitally. Here, we just consider the sound propagation time required by using a directional transmitter to scan. Suppose we require a maximum imaging range of 75 m, and 200×200 beams need to be formed. For the directional transmitter, the time for forming one beam at the range of 75 m will be $2 \times 75/1500 = 0.10$ s, considering a sound speed $c = 1500$ m/s. Then the total time of beamforming computes to $200 \times 200 \times 0.10 = 1.11$ h. We can find that the time required by the directional transmitter cannot meet the real-time requirement. Directional transmitters are prohibited to be used in underwater real-time 3-D acoustical imaging. Therefore, the transmitter should be omnidirectional. To achieve the lateral resolution for underwater real-time 3-D acoustical imaging systems, a 2-D receiving aperture is indispensable.

Two kinds of underwater real-time 3-D acoustical imaging systems: practical and at the simulation stages are introduced in the following.

1.1.1 Practical Systems

A series of real-time Echoscope 3-D imaging sonars have been launched by Coda Octopus company [5–8]. Figure 1.3 shows the dredging images captured by the Echoscope real-time 3-D imaging sonar and an underwater vehicle's optical camera, as reported on Code Octopus' website [9]. Figure 1.4 shows the shipwreck images captured by the Echoscope underwater real-time 3-D imaging sonar and a multi-beam sonar [10]. It can be seen from Fig. 1.3 that for underwater engineering operations, real-time 3-D acoustical imaging provides more useful information than underwater

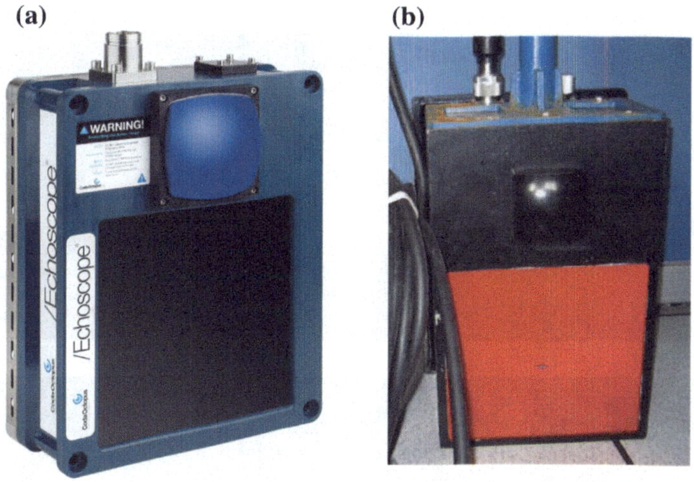

Fig. 1.2 Wet ends of Echoscope 3-D real-time system (a) and the system (b) in [2]

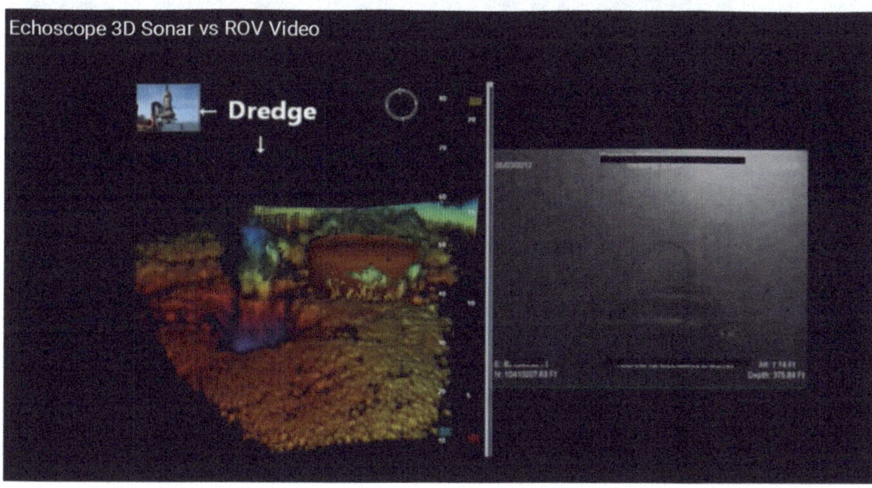

Fig. 1.3 Images obtained by Echoscope real-time 3-D imaging sonar and ROV's optical camera [8]

optical videos. Compared to multi-beam sonars, underwater real-time 3-D acoustical imaging system delivers better imaging quality in Fig. 1.4. Figures 1.3 and 1.4 demonstrate that for many underwater applications such as underwater construction, dredging, underwater archaeology, port and harbor security, and infrastructure inspection, underwater real-time 3-D acoustical imaging is superior.

The Echoscope 3-D imaging sonar work in dual frequencies of 375 kHz and 610 kHz [5]. For the models of 375 kHz and 610 kHz, the maximum imaging ranges are 120 m and 80 m respectively. Higher frequency gives better angular resolution,

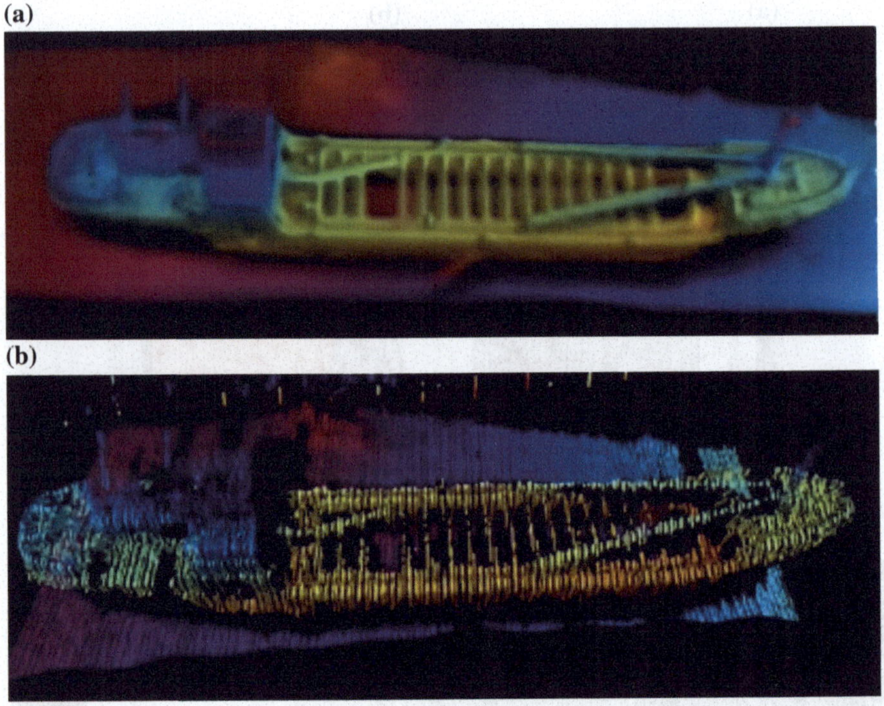

Fig. 1.4 Images obtained by Echoscope real-time 3-D imaging sonar (a) and multi-beam sonar (b) [9]

but the maximum imaging range is decreased because of the frequency-dependent attenuation. Table 1.1 summarizes the specifications of Echoscope 3-D imaging sonar in the model of 375 kHz. The range resolution is 3 cm, which means that the bandwidth is 25 kHz, with a sound speed of 1500 m/s. The working frequency of 375 kHz is much higher than the bandwidth of 25 kHz. The system is hence narrowband. A new real-time 3-D sonar, Echoscope^{4G} Surface [8] was launched by Coda Octopus in January 2018. Compared to its predecessors, this new system is claimed to be 50% lighter and 40% smaller.

The Institute of Acoustics, Chinese Academy of Sciences and Suzhou Soundtech Oceanic Instrument Company [11, 12] collaborated and developed an underwater real-time 3-D acoustical imaging system, as shown in Fig. 1.5. The specifications of the system are also given in Table 1.1. The working frequency of this system is 300 kHz. The maximum imaging range of the system is 50 m. Figure 1.5 also shows the images captured by the system. It can be seen that the diver and anchor can be clearly visualized by the system in Fig. 1.5.

Supported by the National High-Tech Research and Development of China, Chen et al. developed an underwater real-time 3-D acoustical imaging system shown in Fig. 1.2(a) [2], [13–15]. The working frequency of this system is 300 kHz. The

Table 1.1 Parameters of two typical real-time 3-D acoustical imaging systems: Echoscope [9], and the system in [10]

	Echoscope	The system in [10]
Number of beams	128 × 128	128 × 128
Range resolution	3.0 cm	2.5 cm
Angular coverage	50° × 50°	45° × 45°
Minimum range	1 m	1 m
Maximum range	120 m	50 m
Carrier frequency	375 kHz	300 kHz
Update rate (ping rate)	Up to 12 Hz	Up to 12 Hz

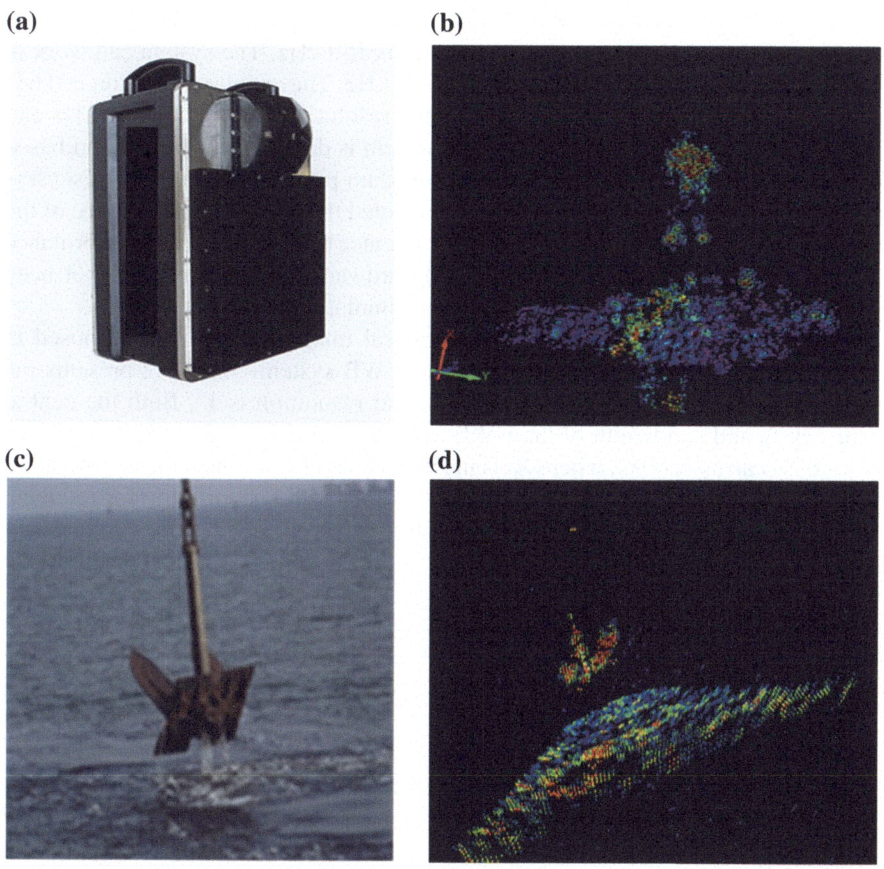

Fig. 1.5 The underwater real-time 3-D acoustical imaging system in [10] and the imaging results. **a** wet end; **b** the diver image obtained; **c** the anchor used in the experiment; **d** the anchor image obtained

system [2] is able to form 128 × 128 beams in real time. Owing to the narrowband beamforming algorithm used [2], the system works in narrowband.

1.1.2 Systems at Simulation Stages

There exist some wideband and ultrawideband (UWB) underwater real-time 3-D acoustical imaging systems currently. However, all of them are at simulation stages. From the existing wideband and UWB systems, we can find the advantages of wideband and UWB underwater real-time 3-D acoustical imaging.

Trucco et al. proposed a system of wideband underwater real-time 3-D acoustical imaging system in [4]. The system bandwidth is 150 kHz. The system can work in the dual central frequencies: 600 kHz and 1200 kHz. The angular resolution is 0.64° at 600 kHz and 0.32° at 1.2 MHz. The range resolution is 5 mm at 600 kHz and 2.5 mm at 1.2 MHz. The 2-D array in the system is designed by the method based on simulated annealing [18] and narrowband beam pattern. The number of sensors of the designed 2-D array is 584. It should be noted that the best performance of the designed 2-D array in narrowband does not guarantee to achieve the best performance in wideband. The experimental results and hardware development have not been reported. The system in [4] only shows some simulation results.

A UWB underwater real-time 3-D acoustical imaging systems is proposed in [16]. Figure 1.6 shows the schematic of the UWB system. Table 1.2 presents the specifications of the UWB system. The angular resolution is 1°. Both the central frequency f_0 and bandwidth of the UWB system are 300 kHz. The aperture size of the used 2-D array is $60\lambda_0$, where λ_0 is the wavelength of f_0. The number of sensors of the 2-D array is only 32, which provides a remarkable opportunity for reducing hardware cost. The system is referred to as ultrasparse. However, similar to [4], only theoretical and simulation results are given in [16]. The UWB system is still at the simulation stage.

A UWB system with an ultralarge ultrasparse 2-D array is proposed in [17], to obtain the high angular resolution of 0.1°. The central frequency f_0 is 300 kHz. The bandwidth of the UWB system is 210 kHz. The 'ultralarge' means that the aperture

Table 1.2 Main features of the prototype of the UWB underwater 3-D acoustical imaging system

Central frequency		300 kHz
Bandwidth		300 kHz
Maximum imaging range		200 m
Number of elements		32
Angular resolution		1°
Field of view	Near field	26° × 26°
	Far field	60° × 60°
Range resolution		2.5 mm

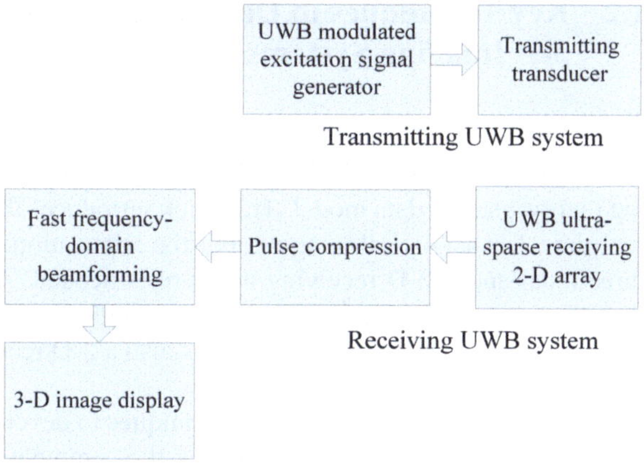

Fig. 1.6 Schematic of the UWB underwater real-time 3-D acoustical imaging system [16]

size of the 2-D array used is $600\lambda_0$. The number of sensors of the ultralarge 2-D array is only 100, which is what the 'ultrasparse' means. If a uniform 2-D array with the aperture of $600\lambda_0$ and the half-wavelength interelement spacing is used, the number of sensors should be 1200×1200, which is so large that the hardware cannot implement. Thus, the concept of ultralarge ultrasparse UWB 2-D array is promising. The UWB system makes it possible to obtain the high angular resolution of 0.1° at a very low hardware cost. Similarly, only theoretical and simulation results are provided to validate the UWB system in [17].

1.1.3 Summary of the Systems

To date, all the existing practical underwater real-time 3-D acoustical imaging systems are narrowband. The advantages of the narrowband systems are low computational load in beamforming and simple hardware implementation. The disadvantages are summarized as: (i) narrowband leads to significant level of speckle noise due to the coherent overlapping of echoes; (ii) range resolution capability is low; (iii) sidelobe levels of narrowband 2-D arrays are high, causing a degradation in imaging quality.

Wideband and UWB real-time 3-D acoustical imaging can weaken the coherent overlapping, improve the range resolution capability and decrease the sidelobe levels of 2-D arrays. In addition, the UWB technique is able to decrease the hardware cost remarkably. Thus, wideband and UWB systems are preferred for developing next-generation underwater real-time 3-D acoustical imaging systems.

1.2 Key Techniques in Developing Underwater Real-Time 3-D Imaging Systems

Before designing and developing underwater real-time 3-D acoustical imaging systems, we should have knowledge about the acoustical propagating and backscattering theory, and the data model. This book introduces the basic theory for underwater real-time 3-D acoustical imaging. For the application, transmitters should be omnidirectional and a 2-D receiving aperture is needed. As analyzed in Chap. 2, even though there exist three 3-D imaging methods: beamforming, acoustic holograph and acoustic lens with a retina which is also a 2-D receiving aperture, beamforming is the most popular method.

Beamforming is one of the key techniques to develop an underwater real-time 3-D acoustical imaging system. However, the computational load of the conventional beamforming methods is overhigh for the real-time 3-D application. To mitigate the computational load, some fast 3-D beamforming methods have been proposed. Different fast beamforming methods have different limitations. For example, some of the fast methods are only suitable for narrowband. This book shows the typical fast 3-D beamforming methods in Chap. 3.

2-D sparse array design is another key technique for underwater real-time 3-D acoustical imaging. According to [1–4], if we want to achieve the angular resolution of 1°, the size of rectangular 2-D arrays employed should be $50\lambda_0 \times 50\lambda_0$. If we use full rectangular 2-D arrays with the half-wavelength interelement spacing, the number of sensors should be 100×100 to obtain the aperture size of $50\lambda_0 \times 50\lambda_0$ [1]. The cost of implementing the hardware 100×100 sensors is exceedingly-high. It is impossible to employ the full 2-D array to develop an underwater real-time 3-D system. To achieve a higher angular resolution, a 2-D array with a bigger size is needed. Thus, sparse design for large 2-D arrays is mandatory for underwater real-time 3-D acoustical imaging. It should be pointed out that the design techniques of narrowband and wideband sparse 2-D arrays are different. The UWB arrays are able to achieve the ultrasparsity to decrease the hardware cost significantly. This book introduces the design techniques of underwater sparse 2-D arrays of narrowband, wideband and UWB respectively in Chap. 4.

Because the cost of developing an underwater real-time 3-D acoustical imaging system is very high, to save the development cost, it is necessary to employ a simulation technique to test the imaging methods and evaluate the performance. The simulation technique for underwater 3-D acoustical imaging is described in Chap. 5.

The principles and steps of designing a underwater real-time acoustical imaging system are explained in Chap. 6. Finally, Chap. 7 proposes some future research potentials for underwater real-time 3-D acoustical imaging.

1.3 Structure of This Book

The rest of this book is organized as follows. Chap. 2 introduces the basic theories of underwater real-time 3-D acoustical imaging. Chap. 3 presents the fast 3-D beamforming methods which can be used in underwater real-time 3-D acoustical systems. Chap. 4 shows the design techniques of large sparse 2-D arrays for the underwater application, including narrowband, wideband and UWB. The simulation technique is shown in Chap. 5. Chapter 6 discusses how to design and implement the underwater real-time 3-D systems. Chapter 7 summarizes the book and shows some future research potentials.

References

1. V. Murino, A. Trucco, Three-dimensional image generation and processing in underwater acoustic vision. Proc. IEEE **88**(12), 1903–1948 (2000)
2. Y. Han, X. Tian, F. Zhou, R. Jiang, Y. Chen, A real-time 3-D underwater acoustical imaging system. IEEE J. Ocean. Eng. **39**(4), 620–629 (2014)
3. X. Liu, F. Zhou, H. Zhou, X. Tan, R. Jiang, Y. Chen, A low complexity real-time 3-D sonar imaging system with a cross array. IEEE J. Ocean. Eng. **41**(2), 262–273 (2016)
4. A. Trucco, M. Palmese, S. Repetto, Devising an affordable sonar system for underwater 3-D vision. IEEE Trans. Instrum. Meas. **57**(10), 2348–2354 (2008)
5. http://www.codaoctopus.com/products/echoscope
6. R.K. Hansen et al., Mosaicing of 3D sonar data sets-techniques and applications, in Proceedings of IEEE/MTS OCEANS Conference, September (2005)
7. A. Davis, A. Lugsdin, High speed underwater inspection for port and harbour security using Coda Echoscope 3D sonar, in Proceedings of IEEE/MTS OCEANS Conf*erence*, September (2005)
8. Coda Octopus Launches Next-Generation Real-Time 3D Sonar. Coda Octopus Group, Inc., Jan (2018)
9. http://www.codaoctopus.com/echoscope-3d-sonar-vs-rov-video-camera-courtesy-fugro-chance-inc-wwwfugrochancecom
10. https://www.youtube.com/watch?v=2d1r2bjibCE
11. http://www.sz-soundtech.com/product/chengxiang/2014-04-28/18.html
12. P. Wang, Y. Ren, Y. Huang, J. Liu, Design and implementation of 3D acoustical imaging sonar signal processing method based on TMS 320C6678. J. Naval Univ. Eng. **15**(2), 85–90 (2014)
13. P. Chen, X. Tian, Y. Chen, Optimization of the digital near-field beamforming for underwater 3-D sonar imaging system. IEEE Trans. Instrum. Meas. **59**(2), 415–424 (2010)
14. L. Yuan, R. Jiang, Y. Chen, Gain and phase autocalibration of large uniform rectangular arrays for underwater 3-D sonar imaging systems. IEEE J. Ocean. Eng. **39**(3), 458–471 (2014)
15. X. Liu, F. Zhou, H. Zhou, X. Tan, R. Jiang, Y. Chen, A low complexity real-time 3-D sonar imaging system with a cross array. IEEE J. Ocean. Eng. **41**(2), 262–273 (2016)
16. C. Chi, Z. Li, Q. Li, Ultrawideband underwater real-time 3-D acoustical imaging with ultra-sparse arrays. IEEE J. Ocean. Eng. **42**(1), 97–108 (2017)
17. C. Chi, Z. Li, Q. Li, High-resolution real-time underwater 3-D acoustical imaging through designing ultralarge ultrasparse ultra-wideband 2-D arrays. IEEE Trans. Instrum. Meas. **66**(10), 2647–2657 (2017)
18. A. Trucco, Thinning and weighting of large planar arrays by simulated annealing. IEEE Trans. Ultrason. Ferroelectr. Freq. Control **46**(2), 347–355 (1999)

Chapter 2
Basic Theory for Underwater Real-Time 3-D Acoustical Imaging

Abstract The data model for underwater real-time three-dimensional (3-D) acoustical imaging is introduced. Three real-time 3-D imaging methods: beamforming, acoustic lens and holograph are analyzed. This chapter points out that beamforming is the most promising method for developing underwater real-time 3-D acoustical imaging systems.

Keywords Acoustic holography · Beamforming · Data model · Generation of 3-D acoustical images

2.1 Data Model for Underwater Real-Time 3-D Imaging

As analyzed in Chap. 1, the transmitters of underwater real-time 3-D acoustical imaging systems are omnidirectional. The transmitted acoustic signal is reflected if it encounters a change in the acoustic impedance. The acoustic impedance is defined as the product of the medium density and acoustic velocity. Typical underwater solid objects, which are man-made or of natural origin, have an acoustic impedance that is very different from that of the water in which they are immersed [1]. Consequently, most portion of the acoustic energy impinging on an underwater object is reflected or scattered outside [1, 2]. Only a small portion of the acoustic energy is transmitted inside the underwater object. The above reveals that underwater acoustical imaging is different from medical ultrasound imaging, where the small discrepancies among the acoustic impedances of the different layers of human body, allow one to image both the external boundary and the internal structure of an organ [1, 3].

As given in [1] and [4–6], the most powerful method for computing the field backscattered by an underwater complex and realistic object is to represent its surface as a collection of point scatterers or small facets, as shown in Fig. 2.1. The method is physically motivated by the Helmholtz-Kirchhoff integral, which is the basis of many theoretical developments associated with scattering [1, 7, 8]. Currently, for underwater real-time 3-D acoustical imaging, the method based on a collection of point scatterers [1, 4, 9] is the most popular to model the echoes of underwater objects, received by a two-dimensional (2-D) aperture.

C. Chi, *Underwater Real-Time 3D Acoustical Imaging*, Signals and Communication Technology, https://doi.org/10.1007/978-981-13-3744-4_2

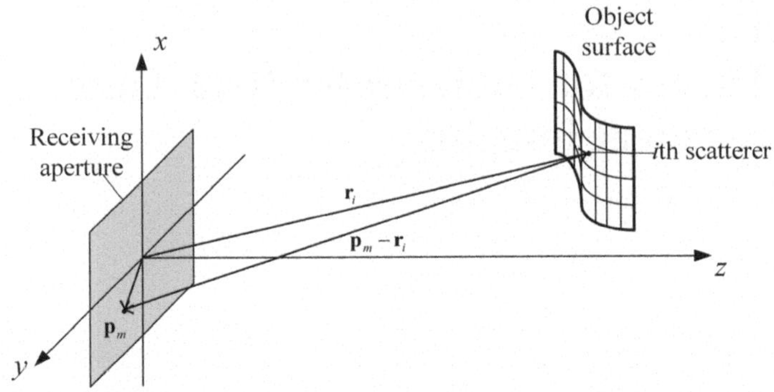

Fig. 2.1 Geometry of the data model of underwater real-time 3-D acoustical imaging

We consider that the surface of an underwater object is composed of Q point scatterers. The ith scatterer is located at the position $\mathbf{r}_i = (x_i, y_i, z_i)$, shown in Fig. 2.1. The distance of the ith scatterer from the coordinate origin is $r_i = |\mathbf{r}_i|$. If an acoustic pulse $p(t)$ is transmitted by an omnidirectional projector in the coordinate origin. In this situation, it is usually assumed that a spherical propagation occurs in an isotropic, linear, absorbing medium [1]. The attenuation of the acoustic pulse is caused by the spherical propagation, frequency-dependent absorption [2, 10] and the characteristics of objects. How these factors influence the received signal will be discussed in Chap. 5. The kth sensor at the 2-D receiving aperture is at the position $\mathbf{p}_m = (x_m, y_m, z_m)$. To explain the theory of underwater 3-D imaging in an easy way, we do not consider the frequency-dependent absorption. The time-domain signal received for the ith scatterer can be written as

$$r(t, \mathbf{p}_m, \mathbf{r}_i) = A_i p(t - \tau_{im}), \tag{2.1}$$

where A_i represents the attenuation caused by the propagation and backscattering, τ_{im} is the propagation delay, expressed as

$$\tau_{im} = \frac{|\mathbf{r}_i|}{c} + \frac{|\mathbf{p}_m - \mathbf{r}_i|}{c}, \tag{2.2}$$

where c is the sound velocity in the medium. The Fourier transform of (2.1) can be written as

$$R(f, \mathbf{p}_m, \mathbf{r}_i) = A_i P(f) \exp(-j2\pi f \tau_{im}), \tag{2.3}$$

where $P(f)$ is the Fourier transform of the transmitted pulse $p(t)$. For the kth sensor, the total received signal from the object surface can be expressed as

$$r(t, \mathbf{p}_m) = \sum_{i=1}^{Q} r(t, \mathbf{p}_m, \mathbf{r}_i). \tag{2.4}$$

The Fourier transform of (2.4) is

$$R(f, \mathbf{p}_m) = \sum_{i=1}^{Q} R(f, \mathbf{p}_m, \mathbf{r}_i). \tag{2.5}$$

Equation (2.4) is taken as the simplest data model of signals received from underwater objects, for underwater real-time 3-D acoustical imaging. In fact, the analysis in the frequency domain is closer to the real situation, but more complex, which will be discussed in Chap. 5.

2.2 Imaging Methods

Underwater real-time 3-D acoustical imaging systems can obtain a 3-D image by processing the signals backscattered by the surfaces of underwater objects. As analyzed in Sect. 2.1, For underwater real-time 3-D imaging, a scene is generally illuminated by an acoustic pulse transmitted by an omnidirectional projector. As summarized in [1], there exist two approaches to process the echoes to generate 3-D images. The first approach is to receive the echoes by a 2-D array of sensors and process the echoes by adequate algorithms: beamforming or holographic, shown in Fig. 2.2. The second approach is to employ an acoustic lens followed by an acoustic retina of sensors, shown in Fig. 2.3.

Beamforming 3-D imaging systems employ a 2-D array of sensors and process the echoes coherently by compensating the delays, to amplify the signal from a predefined direction (steering direction) and suppress all the signals from any other

Fig. 2.2 2-D array of sensors for 3-D beamforming and 3-D acoustic holography

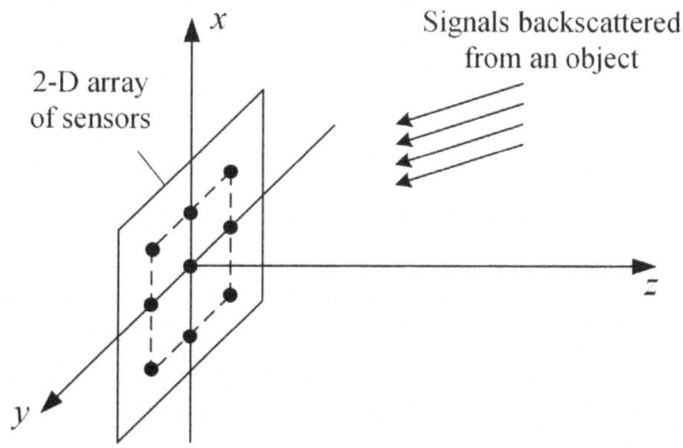

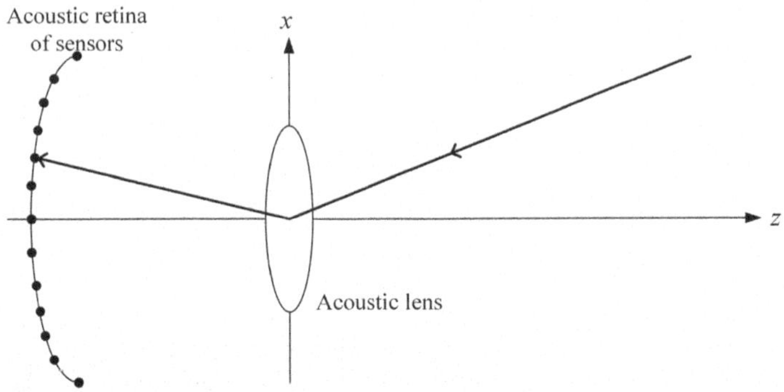

Fig. 2.3 Geometry of a lens-based acoustic imaging system

directions. The output signals of system beamformers provide 3-D information of a scene structure in the steering direction. After the system beamformers scan the whole 3-D scene with beams from many adjacent steering directions, the 3-D image of underwater objects will be reconstructed. Currently, most of the underwater real-time 3-D imaging systems use beamforming to reconstruct 3-D images.

As mentioned in [1] and [11–13], acoustic holographic systems also employ a 2-D array to receive the echoes, and back-propagate the received signals to reconstruct the underwater 3-D structure. Generally, acoustic holography is realized by the inversion of the propagation and scattering equations.

For acoustic lens imaging systems, by using an acoustic lens, backscattered signals are focused on an imaging plane where an acoustic 2-D retina of sensors is placed behind the lens. The signal received by each sensor on the retina is corresponding to a scene response from a predefined direction. Measuring the time-of-flight of an acoustic pulse [1] is easily realizable, which is different from optical imaging systems. Thus, the acoustic lens systems are able to estimate ranges of objects. The retina transforms the acoustic 2-D image frame into electrical signals over time to obtain a 3-D image. Acoustical lens imaging systems require a certain minimum distance between the lens and focusing plane, which results in a large and cumbersome 3-D imaging system. In the last two decades, very few papers have reported on underwater lens 3-D acoustical imaging systems.

2.2.1 Beamforming

Consider a 2-D array with M omnidirectional and point-like sensors, indexed by m. The 2-D array is placed on the plane $z = 0$. The position of a sensor on the 2-D array is denoted by $\mathbf{p}_m$. The received signal of the sensor is denoted by $s_m(t)$. Assume that the unitary vector of the steering beam in the predefined direction is denoted by $\hat{\mathbf{u}}$, the focusing distance is denoted by r_o. The time-domain beam signal, denoted by $b(t, r_0, \hat{\mathbf{u}})$ is [1], [14] expressed as

$$b(t, r_0, \hat{\mathbf{u}}) = \sum_{m=1}^{M} w_m s_m\big(t - \tau(m, r_0, \hat{\mathbf{u}})\big), \tag{2.6}$$

where w_m are the weights to each sensor to suppress the interferences from other directions, $\tau(m, r_0, \hat{\mathbf{u}})$ are the delays required to steer the beam to the direction $\hat{\mathbf{u}}$ at the focusing distance r_0. The expression of $\tau(m, r_0, \hat{\mathbf{u}})$ is given as

$$\begin{aligned}\tau(m, r_0, \hat{\mathbf{u}}) &= \frac{r_0 - |\mathbf{p}_m - r_0\hat{\mathbf{u}}|}{c} \\ &= \frac{r_0 - \sqrt{r_0^2 + |\mathbf{p}_m|^2 - 2r_0\mathbf{p}_m \cdot \hat{\mathbf{u}}}}{c}.\end{aligned} \tag{2.7}$$

The beamforming can also be realized in the frequency domain. The Fourier transform of (2.6) is written as

$$B(f, r_0, \hat{\mathbf{u}}) = \sum_{m=1}^{M} w_m S_m(f) \exp\big(-j2\pi\tau(m, r_0, \hat{\mathbf{u}})\big), \tag{2.8}$$

where $S_m(f)$ is the Fourier transform of $s_m(t)$. Substituting (2.3) and (2.5) into (2.8), we can obtain the following expression:

$$\begin{aligned}B(f, r_0, \hat{\mathbf{u}}) &= \sum_{m=1}^{M} w_m \sum_{i=1}^{Q} A_i P(f) \exp\big[-j2\pi f\big(\tau_{im} + \tau(m, r_0, \hat{\mathbf{u}})\big)\big] \\ &= \sum_{i=1}^{Q} A_i P(f) \sum_{m=1}^{M} w_m \exp\big[-j2\pi f\big(\tau_{im} + \tau(m, r_0, \hat{\mathbf{u}})\big)\big].\end{aligned} \tag{2.9}$$

The beam pattern of the used 2-D array is included in (2.9). Let $BP(r_0, \hat{\mathbf{u}})$ denote the beam pattern. Here, $BP(r_0, \hat{\mathbf{u}})$ can be expressed as

$$BP(r_0, \hat{\mathbf{u}}) = \sum_{m=1}^{M} w_m \exp\big[-j2\pi f\big(\tau_{im} + \tau(m, r_0, \hat{\mathbf{u}})\big)\big]. \tag{2.10}$$

We can see that we can improve the imaging quality by designing a good 2-D array with a well performance in terms of the beam pattern. Thus, the beam pattern of 2-D arrays is important for underwater real-time 3-D acoustical imaging. It should be noted that beam patterns in the far and near fields are different. When designing 2-D arrays, we should pay attention to this. Figure 2.4 shows an far-field beam pattern of a line array. To improve the imaging quality, the sidelobe levels as shown in Fig. 2.4

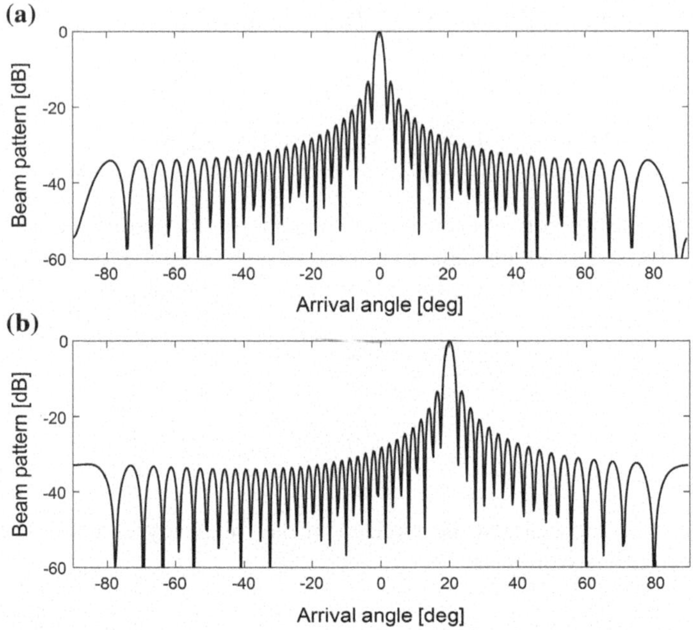

Fig. 2.4 Beam patterns of the line array with 50 sensors, the half-wavelength spacing, and the central frequency of 300 kHz. **a** Steering angle is 0°; **b** Steering angle is 20°

need to be controlled. Detailed discussions about beam patterns will be given in Chap. 4.

Reference [1] points out that some unconventional beamforming methods such as adaptive algorithms will result in excessive computational load and low robustness. To date, no papers have studied adaptive beamforming for underwater real-time 3-D acoustical imaging. Thus, adaptive beamforming methods for underwater real-time 3-D acoustic imaging, will not be discussed in this book.

2.2.2 Acoustic Holography

Without considering the Fresnel approximation, one holographic method in [1] and [15, 16] based on a matrix formulation can be used to generate a 3-D acoustic image. The data model vector received by a 2-D array with M sensors is denoted by $\mathbf{S}(f, \mathbf{p})$, which is an $M \times 1$ column vector. Let $\mathbf{C}(f, \mathbf{r})$ be a $Q' \times 1$ column vector, which represents the reflectivity of each resolution cell contained in the scene volume to be imaged. The number of resolution cells, Q' is different from the number of scatterers, Q used in the data model of (2.5). The ith resolution cell is placed in $\mathbf{r}_i$

($i = 1, \ldots, Q'$). If the ith resolution cell does not contain any object, its reflectivity is null. According to [1] and [15, 16], the data model in (2.5) can be rewritten as

$$\mathbf{S}(f, \mathbf{p}) = \mathbf{U}(f, \mathbf{p}, \mathbf{r})\mathbf{c}(f, \mathbf{r}) \tag{2.11}$$

where is an $M \times Q'$ transfer matrix, the element of which is given as

$$u_{mi} = P(f)\exp\left[\frac{-j2\pi f}{c}(r_i + |\mathbf{p}_m - \mathbf{r}_i|)\right]. \tag{2.12}$$

This holographic method transfers the imaging process into estimating the vector $\mathbf{c}(f, \mathbf{r})$ from the prior knowledge of $\mathbf{U}(f, \mathbf{p}, \mathbf{r})$ and the received vector of $\mathbf{S}(f, \mathbf{p})$. The best estimate of $\mathbf{c}(f, \mathbf{r})$ is expressed as

$$\hat{\mathbf{c}}(f, \mathbf{r}) = \mathbf{U}^H(f, \mathbf{p}, \mathbf{r})\left[\mathbf{U}(f, \mathbf{p}, \mathbf{r})\mathbf{U}^H(f, \mathbf{p}, \mathbf{r})\right]^{+}\mathbf{S}(f, \mathbf{p}), \tag{2.13}$$

where '^{H}' is the operation of complex conjugate and transpose, and '$^{+}$' is the pseudoinverse. For a 3-D imaging system, the resolution cells are generally divided as a sequence of concentric spherical layers. A specific transfer matrix for each layer is defined.

This holographic method requires an offline computation and storing an inversion matrix for each layer and each frequency bin. As analyzed in [1], the amount of storing memory needed is excessively large. In the last 20 years, few papers have discussed this method in underwater real-time 3-D acoustical imaging.

2.2.3 Summary of the Imaging Methods

Acoustical lens imaging system requires a certain minimum distance between the lens and focusing plane, which frequently results in a large and cumbersome 3-D imaging system. The imaging range of some acoustical holographic 3-D imaging systems is inherently small, making this method of imaging less useful for many typical underwater applications [1]. The computational load and memory required by the other holographic algorithm shown in (2.13) are prohibitable for underwater real-time 3-D acoustical imaging [1]. Digital 3-D beamforming imaging system overcomes the disadvantages of the above two acoustical imaging methods, and has been the most popular choice for underwater real-time 3-D imaging.

2.3 Parameters for Underwater Real-Time 3-D Acoustical Imaging Systems

The resolutions of 2-D arrays should be first evaluated for developing underwater 3-D acoustical imaging systems. It is known that the range resolution of the 2-D arrays is determined by the bandwidth of transducers, which is

$$r_{range} = \frac{c}{2\Delta f}, \tag{2.14}$$

where c is the underwater sound velocity, and Δf is the bandwidth of the 2-D arrays. Equation (2.1) shows that the range resolution is in inverse proportion to the bandwidth, which means wideband arrays are necessary for achieving the high range resolution. For a 100% bandwidth 2-D array with 300 kHz central frequency, the range resolution will be 2.5 mm.

It is also known that for a 2-D circular array of the central frequency f_0 with the diameter D and the corresponding wavelength λ_0, the angular resolution θ_{res}, i.e. the main-lobe width of the beam pattern [16], is determined by

$$\theta_{res} \approx \frac{\lambda_0}{D}. \tag{2.15}$$

It means that the larger the diameter D, the higher the angular resolution. Therefore, the angular resolution is substantially decided by the aperture size of arrays, and nearly irrelative to the bandwidth. Based on (2.2), the lateral resolution will be

$$r_{lateral} \approx l\theta_{res}, \tag{2.16}$$

where l is the imaging distance.

2.4 Image Displaying

After doing beamforming for underwater real-time 3-D acoustical imaging, we need to display 3-D image based on the beam outputs. To convert the beam outputs to the images which are easy for human to look at, the interpolation is needed. Figure 2.5 shows the interpolation for displaying a 2-D image. 3-D displaying use the similar operation. When doing the interpolation, there exist some common algorithms [17], such as nearest neighbor, linear and cubic.

This book focuses on how to develop real-time 3-D systems to obtain beam outputs with low cost and high efficiency. The details about the displaying and imaging processing techniques will not be discussed in this book.

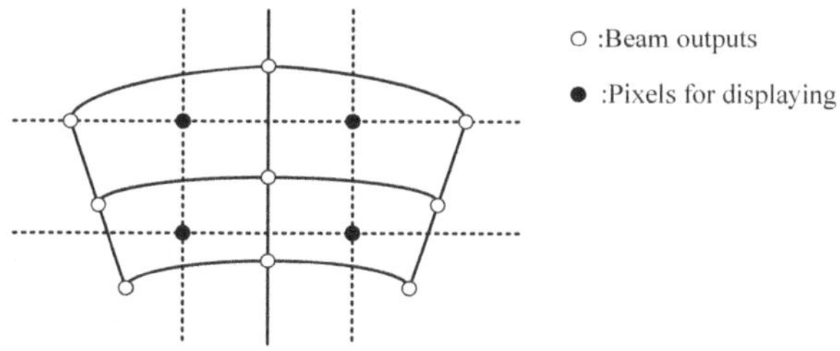

Fig. 2.5 Schematic of image interpolation for displaying of underwater beamforming imaging systems

References

1. V. Murino, A. Trucco, Three-dimensional image generation and processing in underwater acoustic vision. Proc. IEEE **88**(12), 1903–1948 (2000)
2. R.J. Urick, *Principles of Underwater Sound*, 3rd edn. (McGraw-Hill, New York, 1983)
3. Z.H. Cho, J.P. Jones, M. Singh, *Foundations of Medical Imaging* (Wiley, New York, 1993)
4. M. Palmese, A. Trucco, Acoustic imaging of underwater embedded objects: signal simulation for three-dimensional sonar instrumentation. IEEE Trans. Instrum. Meas. **55**(4), 1339–1347 (2006)
5. O. George, R. Bahl, Simulation of backscattering of high frequency sound from complex objects and sand sea-bottom. IEEE J. Oceanic Eng. **20**(2), 119–130 (1995)
6. T.L. Henderson, S.G. Lacker, Seafloor profiling by a wideband sonar: simulation, frequency-response, optimization, and results of a brief sea test. IEEE J. Oceanic Eng. **14**(1), 94–107 (1989)
7. D.E. Funk, K.L. Williams, A physically motivated simulation technique for high-frequency forward scattering derived using specular point theory. J. Acoust. Soc. Amer. **91**(5), 2606–2614 (1992)
8. W.A. Kinney, C.S. Clay, G.A. Sandness, Scattering from a corrugated surface: comparison between experiment, Helmholtz-Kirchhoff theory, and the facet-ensemble method. J. Acoust. Soc. Amer. **73**(1), 183–194 (1993)
9. C. Chi, Z. Li, Q. Li, High-resolution real-time underwater 3-D acoustical imaging through designing ultralarge ultrasparse ultra-wideband 2-D arrays. IEEE Trans. Instrum. Meas. **66**(10), 2647–2657 (2017)
10. R.E. Francois, G.R. Garrison, Sound absorption based on ocean measurements: part I: pure water and magnesium sulfate contributions. J. Acoust. Soc. Amer. **72**(3), 896–907 (1982)
11. J.L. Sutton, Underwater acoustic imaging. Proc. IEEE **67**(4), 554–566 (1979)
12. P.N. Keating, T. Sawatari, G. Zilinskas, Signal processing in acoustic imaging. Proc. IEEE **67**(4), 496–509 (1979)
13. A. Yamani, Three-dimensional imaging using a new synthetic aperture focusing technique. IEEE Trans. Ultrason., Ferroelect., Freq. Contr., **44**(7), 943–947 (1997)
14. M. Palmese, A. Trucco, An efficient digital CZT beamforming design for near-field 3-D sonar imaging. IEEE J. Ocean. Eng. **35**(3), 584–594 (2010)
15. J. C. Bu, C. J. M. van Ruiten, L. F. van der Wal, Underwater acoustical imaging algorithms. In *Proceedings of European Conference on Underwater Acoustics*, Luxembourg, Belgium, pp. 717-720 (1992)
16. R.K. Hansen, P.A. Andersen, 3D acoustic camera for underwater imaging, in Acoustical Imaging, vol. 20, eds. by Y. Wei, B. Gu (Plenum, New York, 1993) pp. 723–727
17. C. De Boor, *A practical guide to splines* (Springer, New York, 1978)

Chapter 3
Fast 3-D Beamforming Methods

Abstract This chapter describes fast beamforming methods for underwater real-time three-dimensional (3-D) acoustical imaging. Currently underwater real-time 3-D acoustical imaging algorithms focus on fast 3-D beamforming. The basic theory of conventional beamforming methods is described first. Then, the general techniques: dynamic focusing and partial overlapping for underwater real-time 3-D beamforming systems are introduced. Three typical fast beamforming methods: time-domain fast Fourier transform (FFT), chirp zeta transform (CZT), nonuniform fast Fourier transform (NUFFT) are shown. The time-domain FFT beamforming method is only suitable for narrowband uniform two-dimensional (2-D) arrays. Both the CZT and NUFFT beamforming methods is wideband, and work in the frequency domain. The CZT beamforming method is accurate, but only suitable for uniform 2-D arrays or sparse arrays thinned from uniform 2-D arrays. The NUFFT beamforming method is approximate, but suitable for arbitrary 2-D arrays. It is proven that the computation errors of the NUFFT beamforming are neglected for underwater 3-D imaging. In most cases, the computational load of the NUFFT beamforming method is lower than that of the CZT beamforming method.

Keywords Chirp zeta transform · Fast fourier transform · Narrowband beamforming · Nonuniform fast fourier transform · Real-time processing · 3-D imaging algorithm · Wideband beamforming

3.1 Basic Beamforming Theory

The coordinate geometry shown in Fig. 3.1a is preferred in underwater real-time 3-D acoustical imaging [1–7], which is different from the commonly-used spherical coordinate geometry shown in Fig. 3.2b. According to the notation in Fig. 3.1a, the unit vector of the steering direction $\hat{\mathbf{u}}$ of a 2-D array can be expressed as

$$\hat{\mathbf{u}} = \left(\sin\alpha, \sin\beta, \sqrt{\cos\alpha^2 - \sin\beta^2}\right), \tag{3.1}$$

C. Chi, *Underwater Real-Time 3D Acoustical Imaging*, Signals and Communication Technology, https://doi.org/10.1007/978-981-13-3744-4_3

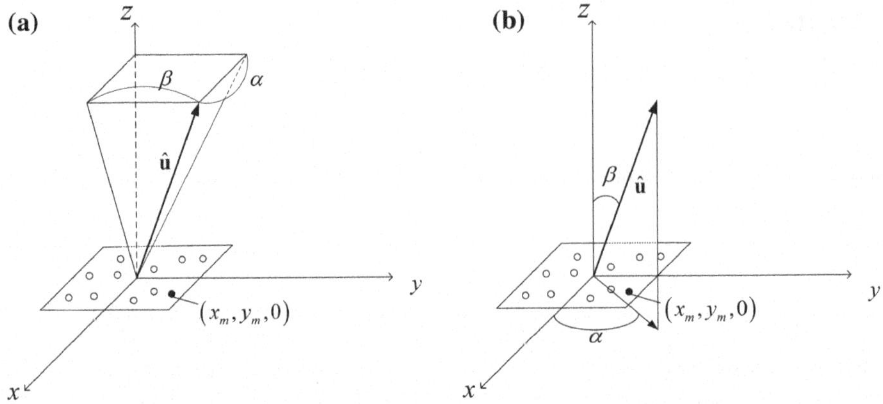

Fig. 3.1 Coordinate geometry preferred in underwater real-time 3-D acoustical imaging (a) and spherical coordinate geometry (b)

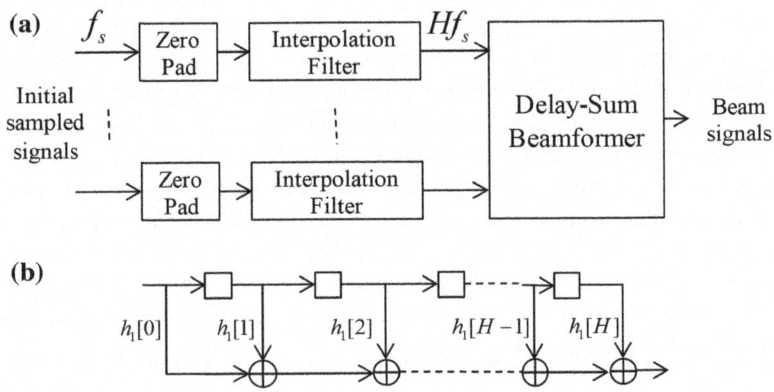

Fig. 3.2 The conventional delay-and-sum beamformer (a) and the structure of the interpolation filter (b)

where α is the angle between the vector $\hat{\mathbf{u}}$ and its projection on the plane yz, and β is the angle between the vector $\hat{\mathbf{u}}$ and its projection on the plane xz.

It is useful to recall that for a uniform 2-D rectangular array, the number of sensors is often given by $M_0 \times N_0$ and the sensor is identified by the indexes (m, n), which is suitable for the time-domain FFT beamforming [2] and the CZT beamforming [3–5]. However, for nonuniform 2-D arrays, the number of sensors cannot be denoted by $M_0 \times N_0$. Thus, in most of the cases described in this chapter, the number of sensors is denoted by M and the sensors are identified by the index m. When describing the CZT beamforming [3–5], $M_0 \times N_0$ is used.

3.1.1 Time-Domain Delay-and-Sum Beamforming

We consider a 2-D array with a rectangular aperture on the plane of $z = 0$ and M elements arbitrarily distributed. Let the system generate $N_b \times M_b$ beams and beam signals indexed by (p, q). For the convenience of description and algorithmic implementation, N_b and M_b are considered to be even. The beam signal in the steering direction (α_p, β_q), focused at a distance r_0, can be written as [3–6]

$$b(r_0, t, \alpha_p, \beta_q) = \sum_{m=1}^{M} w_m s_m\big(t - \tau(r_0, m, \alpha_p, \beta_q)\big), \tag{3.2}$$

where $s_m(t)$ is the backscattered signal received by the mth sensor, w_m is the weighting value of the mth sensor, and $\tau(r_0, m, \alpha_p, \beta_q)$ represents the delays required to steer the beam to the direction (α_p, β_q) at the focusing distance r_0. The expression of $\tau(r_0, m, \alpha_p, \beta_q)$ is [6, 8]

$$\begin{aligned}\tau(r_0, m, \alpha_p, \beta_q) &= \frac{r_0 - |\mathbf{v}_m - r_0\hat{\mathbf{u}}|}{c} \\ &= \frac{r_0 - \sqrt{r_0^2 + x_m^2 + y_m^2 - 2x_m r_0 \sin\alpha_p - 2y_m r_0 \sin\beta_q}}{c},\end{aligned} \tag{3.3}$$

where $\mathbf{v}_m$ is the position vector of the mth sensor, $(x_m, y_m, 0)$. $\tau(r_0, m, \alpha_p, \beta_q)$ is utilized to compensate for the propagation time differences from the signal source to the individual array sensors. By using (3.2), the time-domain signals from the direction $\hat{\mathbf{u}}$ and the distance r_0 are amplified coherently, while those from other distances and directions are suppressed. For the uniform 2-D rectangular array with $M_0 \times N_0$ elements, the element position vector is denoted by $\mathbf{v}_{m,n} = (x_m, y_n, 0)$,. Correspondingly, the beam signal of (3.2) can be rewritten as

$$b(r_0, t, \alpha_p, \beta_q) = \sum_{m=1}^{M_0} \sum_{n=1}^{N_0} w_{m,n} s_{m,n}\big(t - \tau(r_0, m, n, \alpha_p, \beta_q)\big), \tag{3.4}$$

where $w_{m,n}$ is the weighting value of the sensor indexed by (m, n), $s_{m,n}(t)$ is the corresponding received signal, and $\tau(r_0, m, n, \alpha_p, \beta_q)$ is the required delay. Similar to (3.3), the expression of the required delay $\tau(r_0, m, n, \alpha_p, \beta_q)$ is

$$\begin{aligned}\tau(r_0, m, n, \alpha_p, \beta_q) &= \frac{r_0 - |\mathbf{v}_{m,n} - r_0\hat{\mathbf{u}}|}{c} \\ &= \frac{r_0 - \sqrt{r_0^2 + x_m^2 + y_n^2 - 2x_m r_0 \sin\alpha_p - 2y_n r_0 \sin\beta_q}}{c}.\end{aligned} \tag{3.5}$$

Equations (3.2) and (3.4) explain the conventional beamforming in the time domain, which is referred to as the conventional delay-and-sum beamforming [3, 8, 9]. The conventional delay-and-sum beamforming is often taken as the baseline reference. The schematic of the conventional delay-and-sum beaforming is shown in Fig. 3.2a. Oftentimes, the sampling frequency f_s of the acoustic systems is not high enough for using the conventional delay-and-sum beamforming. Thus, a finite-impulse response interpolation filter with H stages shown in Fig. 3.2b is adopted to obtain more accurate delays. The computational load of the conventional delay-and-sum beamforming is evaluated by the number of on-line real operations including both real additions and real multiplications. For the arbitrary 2-D array, the number of real operations [3, 4, 6] needed by the delay-and-sum beamforming to generate N_b^2 beams is

$$TD_1 = \left[M(H + N_b^2) + M(H + 1)\right] f_s. \tag{3.6}$$

3.1.2 Frequency-Domain Direct Beamforming

If we segment the received signals into blocks with a certain length L, the discrete Fourier transform (DFT) coefficients $B(r_0, l, \alpha_p, \beta_q)$ of the beam signal $b(r_0, t, \alpha_p, \beta_q)$ can be written as

$$B(r_0, l, \alpha_p, \beta_q) = \sum_{m=1}^{M} w_m S_m(l) \exp\left(-i2\pi f_l \tau(r_0, m, \alpha_p, \beta_q)\right), \tag{3.7}$$

where $S_m(l)$ is the lth DFT coefficient of $s_m(t)$, and the discrete frequency f_l is

$$f_l = l f_s / L, \tag{3.8}$$

where f_s is the sampling frequency and l is the frequency index, $l \in [0, L)$. The DFT can be fast realized by the FFT. When we consider the uniform 2-D rectangular array, the DFT coefficients can be similarly written as

$$B(r_0, l, \alpha_p, \beta_q) = \sum_{m=1}^{M_0} \sum_{n=1}^{N_0} w_{m,n} S_{m,n}(l) \exp\left(-i2\pi f_l \tau(r_0, m, n, \alpha_p, \beta_q)\right), \tag{3.9}$$

Equations (3.7) and (3.9) shows the conventional frequency-domain beamforming, which is usually called as the frequency-domain direct method in many references [3–6, 11–13]. The frequency-domain direct method interprets (3.7) and (3.9) as a complex dot directly in both far and near fields. The fast beamforming such as CZT and NUFFT, all focus on how to realize fast computation of (3.7) or (3.9). For the arbitrary 2-D array, the number of real operations needed by the frequency-domain direct method to generate N_b^2 beams is

$$F_{DM} = (8M - 2)N_b^2. \tag{3.10}$$

To obtain time-domain beam outputs, the inverse discrete Fourier transform (IDFT) of $B(r_0, l, \alpha_p, \beta_q)$ is needed. The IDFT can be fast realized by the inverse fast Fourier transform (IFFT).

To sum up, the frequency-domain direct method is realized by the following three steps:

(1) Input FFT: segment the received broadband signals into sequential blocks and convert the blocks into frequency-domain signals by using FFT (partial overlapping needed [4]).
(2) Spatial processing: interpret (3.7) and (3.9) as a complex dot directly in both far and near fields, to compute beam signals at every valid frequency point.
(3) Output IFFT: invert the beam signals in the frequency domain into the time-domain signals by using IFFT.

3.1.3 Delay Approximation

The delay approximations for 3-D beamforming in both the far and near fields are introduced in the following. Consider that D is the side size of a rectangular 2-D array, or the diameter of a circular 2-D array. The far-field approximation condition [3, 4, 14, 15] is $r_0 > D^2/2\lambda$. The approximation delay [3, 4] in the far field can be expressed as

$$\tau(r_0, m, \alpha_p, \beta_q) = \frac{x_m \sin\alpha_p + y_m \sin\beta_q}{c}. \tag{3.11}$$

For the near-field condition, the Fresnel zone $(0.96D < r_0 \le D^2/2\lambda)$ is the main concern. The Fresnel delay approximation [3, 8, 14] is often adopted in most of the fast beamforming techniques. The approximation delay in the near field can be given by [3]

$$\tau(r_0, m, \alpha_p, \beta_q) = \frac{x_m \sin\alpha_p + y_m \sin\beta_q}{c} - \frac{x_m^2 + y_m^2}{2r_0 c}. \tag{3.12}$$

For the uniform 2-D array, the delay in the far field can be expressed as

$$\begin{aligned}\tau(r_0, m, n, \alpha_p, \beta_q) &= \frac{x_m \sin\alpha_p + y_n \sin\beta_q}{c} \\ &= \frac{md \sin\alpha_p + nd \sin\beta_q}{c},\end{aligned} \tag{3.13}$$

where d is the interelement spacing of the uniform 2-D array. The delay $\tau(r_0, m, n, \alpha_p, \beta_q)$ in the near field can be expressed as

$$\begin{aligned}\tau(r_0, m, n, \alpha_p, \beta_q) &= \frac{x_m \sin\alpha_p + y_n \sin\beta_q}{c} - \frac{x_m^2 + y_n^2}{2r_0c} \\ &= \frac{md \sin\alpha_p + nd \sin\beta_q}{c} - \frac{(md)^2 + (nd)^2}{2r_0c}.\end{aligned} \tag{3.14}$$

3.2 General Techniques for Different Beamforming Methods

In the near field, to mitigate the computational load of beamforming, the dynamic focusing technique [4, 6] is necessary for using all the fast beamforming methods to reconstruct images for underwater real-time 3-D acoustical imaging systems. Underwater real-time 3-D acoustical imaging systems work step-by-step on successive segments of the signals received by the sensors, rather than process all the received signals directly. For most of the beamforming methods, the operation of partial overlapping [4] is necessary for computing the beam signals without distortion. The dynamic focusing and partial overlapping are the two general techniques for underwater real-time 3-D acoustical imaging.

3.2.1 Dynamic Focusing

Under the assumption of the far field, the fast beamforming methods can accelerate 3-D beamforming. Comparing (3.11) to (3.12), the difference is the second term on the right side of (3.12), which includes the focusing distance. If the phase changes caused by the second term on the right side of (3.12), can be compensated before computing beam signals, the fast beamforming methods in the far field can also be applied to the near field. The dynamic focusing is to compensate the phase changes, at the focusing distance r_0. When using the dynamic focusing technique, (3.7) can be written as

$$B(r_0, l, \alpha_p, \beta_q) = \sum_{m=1}^{M} w'_m S_m(l) \exp(-i2\pi f_l \tau(r_0, m, \alpha_p, \beta_q)), \tag{3.15}$$

where w'_m is given as

$$w'_m = w_m \exp\left[j2\pi f\left(\frac{x_m^2 + y_m^2}{2r_0c}\right)\right]. \tag{3.16}$$

The depth of field at the distance r_0 [2, 4, 10] is defined by 3 dB attenuation points in the beam-power along the range. 'Dynamic' means for different distances, the phase compensation of (3.15) should be different. The depth of field is [2, 4] approximated expressed as

$$r_{0+} = r_0 + \frac{r_0}{\left(\frac{D^2}{2\lambda r_0}\right) + 1}, \tag{3.17}$$

$$r_{0-} = r_0 - \frac{r_0}{\left(\frac{D^2}{2\lambda r_0}\right) - 1}. \tag{3.18}$$

3.2.2 *Partial Overlapping*

As pointed out in [4], the partial overlapping of adjacent blocks is necessary for beamforming operations which work block-by-block on the successive segments of the signals received by the sensors.

To compute the beam signal at a given time t_0, we need to sum the samples from all the sensors, referred to different instants for different sensors (before and after t_0), under the delays required by beamforming. Figure 3.3 gives an example. When process a block signals from the sensors, only the central part of the block in the computed beam signal is correct. In the example, the block length is assumed to be 8 and five block signals $s_1, \ldots, s_5$ are delayed and summed to obtain the corresponding block of beam signal. In Fig. 3.3, we can find that only the four central samples of the beam signal block is correct, because all the samples required by computing the four samples are in the current signal blocks. Due to the different delays for different sensors required by steering the beam to a given direction, the two initial and two final samples of the beam signal bock are not correct. If we want to make sure that all the samples of the beam signal block are correct, a four-samples overlapping of the current block with the previous one and the next one is required.

The amount of block overlapping is based on the maximum delay of the sensor signals required by computing all the desired beam signals. The maximum delay is determined by the sensor position, the focusing distance and the steering direction. Generally, the maximum positive delay $\tau_{\max}$ and the minimum negative delay $\tau_{\min}$ can be calculated in advance of system deployment. The number O of samples which should be overlapped for computing all the beam signal samples correctly is given by [4]

$$O = \text{ceil}(f_s \tau_{\max}) - \text{ceil}(f_s \tau_{\min}), \tag{3.19}$$

where the function ceil(a) round a to the nearest integer towards infinity [4].

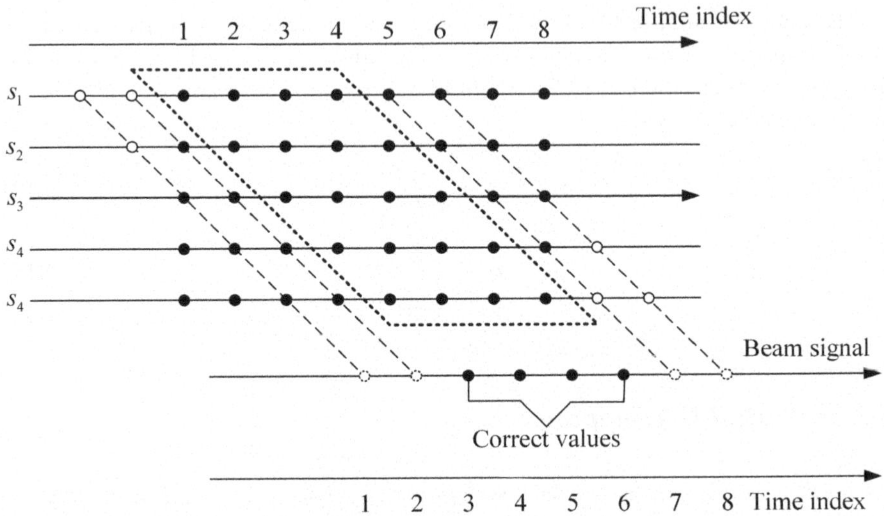

Fig. 3.3 Example of computing a beam signal block, by delaying and summing the signals sampled from the sensors. Five signal blocks $s_1, \ldots, s_5$ are considered. The block length is 8

3.3 Time-Domain FFT Beamforming

The time-domain FFT beamforming method [2] is suitable for uniform 2-D arrays and sparse arrays thinned from uniform 2-D arrays. The following discussion is done in the far field and based on (3.4) and (3.13). In the near field, the dynamic technique can be used for applying the time-domain FFT beamforming. We assume that $s_{m,n}(t)$ received by the sensor indexed by (m, n) is narrowband and complex. Under this assumption, $s_{m,n}(t)$ can be written as

$$s_{m,n}(t) = A_{m,n}(t)\exp(j2\pi f_0 t), \tag{3.20}$$

where $A_{m,n}(t)$ is the signal complex envelop, and f_0 is the carrier frequency. Equation (3.4) can be rewritten as

$$\begin{aligned} b(r_0, t, \alpha_p, \beta_q) &= \sum_{m=1}^{M_0}\sum_{n=1}^{N_0} w_{m,n}A_{m,n}(t)\exp\left[j2\pi f_0\left(t - \tau(r_0, m, n, \alpha_p, \beta_q)\right)\right] \\ &= \exp(j2\pi f_0 t)\sum_{m=1}^{M_0}\sum_{n=1}^{N_0} w_{m,n}A_{m,n}(t)\exp\left[-j2\pi f_0\tau(r_0, m, n, \alpha_p, \beta_q)\right]. \end{aligned} \tag{3.21}$$

Substituting (3.13) into (3.21) and demodulating the carrier frequency f_0, (3.21) can be rewritten as

$$b(t, \alpha_p, \beta_q) = \sum_{m=1}^{M_0} \sum_{n=1}^{N_0} w_{m,n} A_{m,n}(t) \exp\left[j2\pi f_0 \left(-\frac{md \sin \alpha_p + nd \sin \beta_q}{c} \right) \right]. \tag{3.22}$$

Let $\sin \alpha_p$ be defined as

$$\alpha_p = \arcsin \frac{pc}{d \cdot f_0 \cdot M_0}, \quad -\frac{M_0}{2} \leq p \leq \frac{M_0}{2} - 1. \tag{3.23}$$

Let $\sin \beta_q$ be defined as

$$\beta_q = \arcsin \frac{qc}{d \cdot f_0 \cdot N_0}, \quad -\frac{N_0}{2} \leq q \leq \frac{N_0}{2} - 1. \tag{3.24}$$

Considering (3.23) and (3.24), (3.22) can be written as

$$b(t, \alpha_p, \beta_q) = \sum_{m=1}^{M_0} \sum_{n=1}^{N_0} w_{m,n} A_{m,n}(t) \exp\left(-j2\pi \frac{pm}{M_0} \right) \exp\left(-j2\pi \frac{qn}{N_0} \right). \tag{3.25}$$

We can find that (3.25) is the expression of the 2-D discrete Fourier transform. Thus, the fast computation of (3.25) can be realized by the 2-D FFT, which is referred to as the time-domain FFT beamforming method. The steps of the time-domain FFT beamforming method are summarized as follows:

(i) Demodulate the sampled signals from the sensors to obtain the complex envelopes;
(ii) Perform the 2-D FFT at each sample of the complex envelopes to realize 3-D beamforming;
(iii) Apply the dynamic focusing technique before performing the 2-D FFT in the near field.

To evaluate the computational load, we assume $M_0 = N_0$. For the direct computation of (3.22), namely phase-shift beamforming, to generate $N_0 \times N_0$ beams, N_0^4 complex multiplications are needed [2]. For the time-domain FFT beamforming, only $N_0^2 \log N_0$ complex multiplications are needed. It can be seen that the time-domain FFT beamforming is much more efficient than the direct phase-shift beamforming. However, it should be noted that the method is narrowband and only suitable for uniform 2-D arrays or sparse arrays thinned from uniform 2-D arrays.

3.4 CZT Beamforming

The CZT 3-D beamforming is a promising method for underwater real-time 3-D acoustical imaging. Before introducing the CZT beamforming, it is necessary to predefine the steering angles. The CZT beamforming method also employs the coordinate notation shown in Fig. 3.1a. Assume that α_i and α_f are the initial and final azimuth angles, and β_i and β_f represent the initial and final elevation angles. Generally, there are cases where $\alpha_f = -\alpha_i$ and $\beta_f = -\beta_i$. For underwater real-time 3-D acoustical imaging, the predefined angles α_p and β_q are usually equispaced in the sine domain [3–6]. Considering to form $M_b \times N_b$ beams, the beam spacings are decided by

$$\Delta s_\alpha = \frac{\sin\alpha_f - \sin\alpha_i}{M_b - 1}, \tag{3.26}$$

and

$$\Delta s_\beta = \frac{\sin\beta_f - \sin\beta_i}{N_b - 1}. \tag{3.27}$$

Technical details about the CZT 3-D beamforming [3, 4] in the far field is given as follows. For the CZT 3-D beamforming, an uniform 2-D array or sparse array thinned from an uniform 2-D array is mandatory. In the far field, according to (3.13) and the index (m, n), (3.9) can be rewritten as

$$\begin{aligned} B\big(l, \alpha_p, \beta_q\big) &= \sum_{m=1}^{M_0}\sum_{n=1}^{N_0} w_{m,n} S_{m,n}(l) \exp\left(-i2\pi f_l\left(\frac{md\sin\alpha_p}{c} + \frac{nd\sin\beta_q}{c}\right)\right) \\ &= \sum_{m=1}^{M_0}\sum_{n=1}^{N_0} w_{m,n} S_{m,n}(l) z_{\alpha_p}^m z_{\beta_q}^n \end{aligned} \tag{3.28}$$

where

$$z_{\alpha_p} = \exp\left(-i2\pi f_l \frac{d\sin\alpha_p}{c}\right), \tag{3.29}$$

$$z_{\beta_q} = \exp\left(-i2\pi f_l \frac{d\sin\beta_q}{c}\right). \tag{3.30}$$

Equation (3.28) can be regarded as a complex polynomial in $z_{\alpha_p} z_{\beta_q}$ [3, 4]. Based on (3.26) and (3.27), the predefined steering angles, which are equispaced in the sine domain, are expressed as

$$\sin\alpha_p = \sin\alpha_i + p\Delta s_\alpha, \quad (p = 0, \ldots, M_b - 1) \tag{3.31}$$

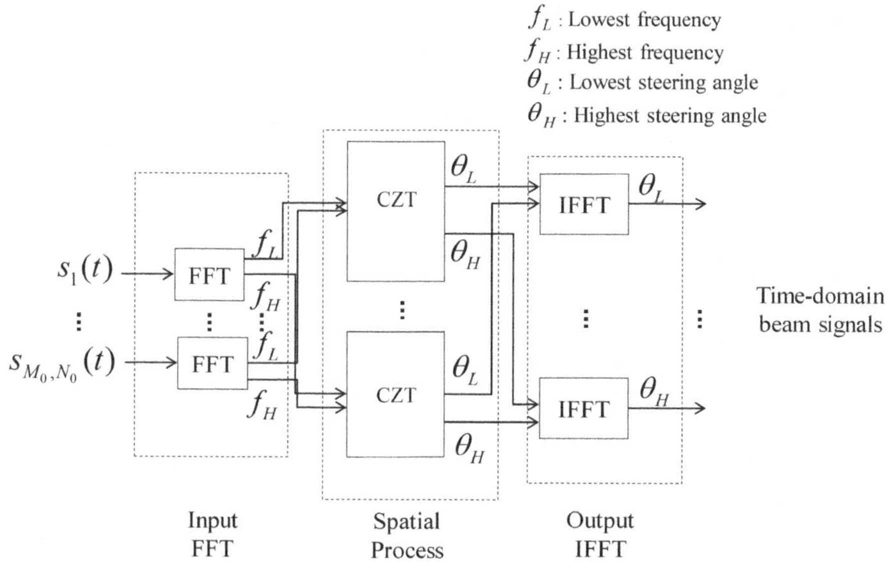

Fig. 3.4 Schematic of the CZT 3-D beamforming method

$$\sin\beta_q = \sin\beta_i + q\Delta s_\beta, \quad (q = 0, \ldots, N_b - 1). \tag{3.32}$$

Under the predefined angles of (3.31) and (3.32), the CZT algorithm can be exploited to fast compute (3.28) [3]. By controlling the values of $w_{m,n}$, the CZT 3-D beamforming can be applied to the sparse arrays thinned from uniform 2-D arrays. The schematic of the CZT 3-D beamforming method is shown in Fig. 3.4. In the near field, the dynamic focusing technique should be employed before using the CZT to do beamforming.

The CZT algorithm is introduced in the following. We can rewrite (3.28) as

$$B(l, \alpha_p, \beta_q) = W_a^{\frac{-p^2}{2}} W_e^{\frac{-q^2}{2}} \sum_{m=0}^{M_0-1} \sum_{n=0}^{N_0-1} w_{m,n} S_{m,n}(l) A_a^m A_e^n W_a^{\frac{-m^2}{2}} W_e^{\frac{-n^2}{2}} \times W_a^{\frac{-(p-m)^2}{2}} W_e^{\frac{-(q-n)^2}{2}} \tag{3.33}$$

where

$$\begin{cases} A_a = \exp\left(-\frac{j2\pi f_l d \sin\theta_{ai}}{c}\right) \\ W_a = \exp\left(\frac{j2\pi f_l d \Delta s_a}{c}\right) \end{cases} \tag{3.34}$$

and

$$\begin{cases} A_e = \exp\left(-\frac{j2\pi f_l d \sin\theta_{ei}}{c}\right) \\ W_e = \exp\left(\frac{j2\pi f_l d \Delta s_e}{c}\right) \end{cases}. \tag{3.35}$$

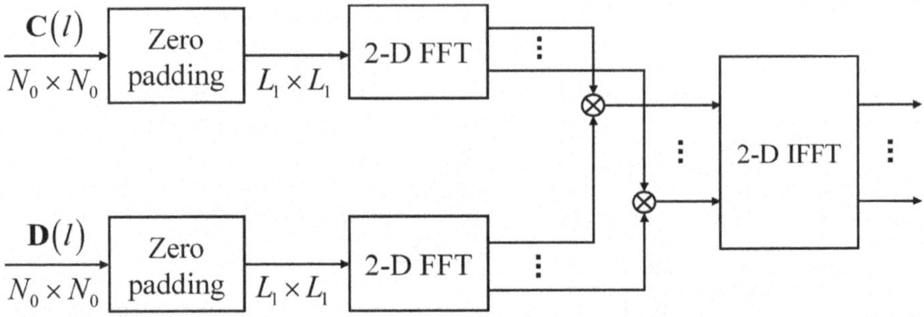

Fig. 3.5 Schematic of the fast convolution with the 2-D FFT and IFFT

Consider the two matrices $\mathbf{C}(l)$ and $\mathbf{D}(l)$. The elements of the two matrixes are respectively expressed as

$$\begin{cases} C_{m,n}(l) = w_{m,n} S_{m,n}(l) A_a^m A_e^n W_a^{\frac{-m^2}{2}} W_e^{\frac{-n^2}{2}} \\ D_{m,n}(l) = W_a^{\frac{m^2}{2}} W_e^{\frac{n^2}{2}} \end{cases}. \tag{3.36}$$

Equation (3.33) transfer $B(l, \alpha_p, \beta_q)$ as a 2-D discrete convolution of $\mathbf{C}(l)$ and $\mathbf{D}(l)$. The 2-D discrete convolution can be realized by the 2-D FFT [3, 16]. To apply the 2-D FFT, the matrices $\mathbf{C}(l)$ and $\mathbf{D}(l)$ need to be zero padded. Consider $M_0 = N_0$ and $M_b = N_b$. The matrices $\mathbf{C}(l)$ and $\mathbf{D}(l)$ have a size $N_0 \times N_0$. The output beam matrix should have a size $N_b \times N_b$. To prevent wraparound from contaminating the computation of the linear convolution [4], the matrix should have a size $L_1 \times L_1$, where $L_1 \geq N_0 + N_b - 1$. Generally, L_1 should be a power of two to fully exploit the FFT advantage. The FFT implementation of such a convolution is performed by the following steps:

(1) zero pad $\mathbf{C}(l)$ and $\mathbf{D}(l)$ are to obtain a size $L_1 \times L_1$;
(2) Perform the 2-D FFTs of both the two zero-padded matrices;
(3) Multiply aach coefficient of the first FFT by the corresponding coefficient of the second FFT, requiring $L_1 \times L_1$ complex multiplications;
(4) Perform the 2-D IFFT of the results of step (3) to obtain the outputs of the convolution (Fig. 3.5).

To generate N_b^2 beams, the spatial processing of the CZT beamforming method at a single frequency f_l requires the following number of on-line real operations [3, 6]:

$$F_{Csp} = 6\left[N_0^2 + N_b^2 + L_1^2\right] + 20L_1^2 \log_2(L_1). \tag{3.37}$$

The specific evaluation of the CZT 3-D beamforming method will be shown in Sect. 3.6.

3.5 NUFFT 3-D Beamforming

The NUFFT 3-D beamforming [6] which is suitable for arbitrary 2-D arrays, is a promising method for underwater real-time 3-D acoustical imaging. The computational load of the NUFFT 3-D beamforming is also lower than that of the CZT 3-D beamforming. The technical details about the NUFFT 3-D beamforming are introduced in the following.

3.5.1 NUFFT

FFT is proposed to improve the speed of DFT. FFT needs nodes in both the frequency and time domain to be uniformly spaced [16]. NUFFT has been proposed to ensure the fast computation of the DFT of nonuniform nodes, with only a slight drop in the accuracy of the computation [17, 18]. In addition, NUFFT has been applied in many fields such as MRI [19], CT [20], through-wall radar [21], ultrasound plane-wave imaging [22], synthesis of large arbitrary arrays [23], and reconstruction of photoacoustic images [24].

Given that x_k ($k = -K/2, \ldots, K/2 - 1$ and K being a positive even number) are equispaced signal samples, the one-dimensional (1-D) NUDFT coefficients of x_k have the form

$$X(\omega_m) = \sum_{k=-K/2}^{K/2-1} x_k \exp(-ik\omega_m), \quad m = 0, \ldots, M-1, \tag{3.38}$$

where M is a natural number, ω_m is the arbitrary frequency nodes, and $\omega_m \in [-\pi, \pi)$ [18]. The proposed method in this paper will employ the adjoint 2-D NUDFT. To make the description easy to understand, we give the definition of the adjoint 1-D NUDFT [17, 18], expressed as

$$\hat{x}_k = \sum_{m=0}^{M-1} X(\omega_m) \exp(ik\omega_m), \quad k = -K/2, \ldots, K/2 - 1, \tag{3.39}$$

where $\hat{x}_k$ s are the equally spaced samples obtained from the adjoint 1-D NUDFT operations, which may not be equal to x_k s because the NUDFT is not always invertible as the IDFT done.

In general, the algorithms of 1-D NUFFT consist of two steps: the oversampled DFT and the interpolation [18]. They can be described by

$$Y_{k'} = \sum_{k=-K/2}^{K/2-1} x_k \exp\left(\frac{-i2\pi kk'}{K_1}\right), \quad k' = -K_1/2, \ldots, K_1/2 - 1, \tag{3.40}$$

and

$$\hat{X}(\omega_m) = \sum_{j=1}^{J} Y_{(k'_m+j)} u_j^*(\omega_m), \quad m = 0, \ldots, M-1, \tag{3.41}$$

where $Y_{k'}$ denotes the oversampled DFT coefficients with K_1 points ($K_1 \geq K$), K_1/K is defined as the oversampled factor, σ and k'_m in (3.41) is given [18] by

$$k'_m = \begin{cases} \left(\arg\min_{k' \in Z} \left|\omega_m - \frac{2\pi k'}{K_1}\right|\right) - \frac{J+1}{2}, & J \text{ odd}, \\ \left(\max\left\{k' \in Z : \omega_m \geq \frac{2\pi k'}{K_1}\right\}\right) - \frac{J}{2}, & J \text{ even}. \end{cases} \tag{3.42}$$

In (3.41), $\hat{X}(\omega_m)$ are the approximated coefficients in the nonuniform frequency points, $u_j(\omega_m)$ are appropriate frequency-domain interpolation coefficients, "*" denotes complex conjugate, J is the largest number of the nearest nonzero neighbors applied by the interpolation, and $J << K$. $u_j(\omega_m)$ can be obtained through the min-max criterion that Fessler and Sutton proposed [18]. Mathematically, the min-max criterion is an optimization problem which has the form of $\min_{\mathrm{u}(\omega_m) \in C^J} \max_{\mathrm{x} \in C^K : \|\mathrm{x}\| \leq 1} \left|\hat{X}(\omega_m) - X(\omega_m)\right|$ $\left(\mathrm{u}(\omega_m) = (u_1(\omega_m), \ldots, u_J(\omega_m)) \text{ and } \mathrm{x} = (x_{-K/2}, \ldots, x_{K/2-1})\right)$ and has an analytical solution as is derived in [18]. It means that for each desired frequency ω_m, the coefficients $u_j(\omega_m)$ minimize the worst-case approximation error $e = \left|\hat{X}(\omega_m) - X(\omega_m)\right|$. σ and J can be determined by the permitted error limit through the empirical formula $e_{\max} \approx 0.75 \exp[-J(0.29 + 1.03 \log \sigma)]$ [18].

The fast algorithms of the adjoint NUDFT, i.e., the adjoint operator of NUFFT, compute (3.38) by "reversing" (not inverting!) the steps of the algorithms of NUFFT [18] when a nonuniform frequency-domain series $X(\omega_m)$ is provided. The first step is the interpolation and the second one is the oversampled IFFT. Therefore, the interpolation of the adjoint operator of NUFFT can be described as

$$\tilde{X}_{k'} = \sum_{m=0}^{M-1} v_{mk'} X(\omega_m), \tag{3.43}$$

which is akin to "gridding" [18]. $v_{mk'}$ are the sparse interpolation coefficients applied for "gridding". The second step is given by

$$\tilde{x}_k = \sum_{k'=-K_1/2}^{K_1/2-1} \tilde{X}_{k'} \exp(i 2\pi \frac{k'k}{K_1}), \tag{3.44}$$

where $\tilde{x}_k$ is an approximation of $\hat{x}_k$ in (3.39). The adjoint operator of the 1-D NUFFT uses the transpose of the sparse interpolation matrix applied by the 1-D NUFFT [18].

Similar with the definition of the 1-D NUDFT, the definition of the 2-D NUDFT [25] is expressed as

$$X(\omega_{1m}, \omega_{2m}) = \sum_{k_1=-K/2}^{K/2-1} \sum_{k_2=-K/2}^{K/2-1} x_{k_1,k_2} \exp[-i(k_1\omega_{1m} + k_2\omega_{2m})] \quad (m = 0, \ldots, M-1), \tag{3.45}$$

where $X(\omega_{1m}, \omega_{2m})$ are the 2-D NUDFT coefficients of a 2-D uniformly sampled series x_{k_1,k_2}, both ω_{1m} and $\omega_{2m} \in [-\pi, \pi)$. The expression of the adjoint 2-D NUDFT [18, 25] is needed:

$$\hat{x}_{k_1,k_2} = \sum_{m=0}^{M-1} X(\omega_{1m}, \omega_{2m}) \exp[i(k_1\omega_{1m} + k_2\omega_{2m})] \quad \begin{pmatrix} k_1 = -K/2, \ldots, K/2-1 \\ k_2 = -K/2, \ldots, K/2-1 \end{pmatrix}, \tag{3.46}$$

where $\hat{x}_{k_1,k_2}$ s are the 2-D equally spaced samples obtained from adjoint 2-D NUDFT operations, which may not be equal to original x_{k_1,k_2} s because the NUDFT is not always invertible as IDFT done.

The proposed beamforming method will make use of the fast implementation algorithm of the adjoint 2-D NUDFT, i.e., the adjoint operator of 2-D NUFFT. Similar to that of the 1-D NUFFT, when a 2-D nonuniform frequency-domain series $X(\omega_{1m}, \omega_{2m})$ is provided, the adjoint operator of the 2-D NUFFT can be described by

$$\tilde{X}_{k_1',k_2'} = \sum_{m=0}^{M-1} v_{m,k_1',k_2'} X(\omega_{1m}, \omega_{2m}), \tag{3.47}$$

which is also similar to "gridding", and

$$\tilde{x}_{k_1,k_2} = \sum_{k_1'=0}^{K_1-1} \sum_{k_2'=0}^{K_2-1} \tilde{X}_{k_2',k_2'} \exp\left[i2\pi\left(\frac{k_2'k_1}{K_1} + \frac{k_2'k_2}{K_2}\right)\right], \quad \begin{pmatrix} k_1 = -K/2, \ldots, K/2-1 \\ k_2 = -K/2, \ldots, K/2-1 \end{pmatrix}, \tag{3.48}$$

where $\tilde{x}_{k_1,k_2}$ is the approximation of $\hat{x}_{k_1,k_2}$ in (3.46), $v_{m,k_1',k_2'}$ are the sparse interpolation coefficients applied for 2-D "gridding". The interpolation coefficients applied in this paper are also determined by the min-max criterion in Fessler and Sutton's paper [18]. The fast computation of (3.48) can be realized by 2-D IFFT. It should be emphasized that the interpolation matrix used by the adjoint operator of the 2-D NUFFT is the transpose of that used by the 2-D NUFFT [18].

The efficiency and accuracy of the NUFFT and its adjoint operator is dependent on the following crucial parameters: the oversampling factor σ and the number of the nearest neighbors, J. In the following, the computational load relative with σ and J will be analyzed.

For the 1-D NUFFT, because (3.40) can be calculated by FFT with K_1 being a power of two, the amount of real operations required by the oversampled DFT [16, 18] will be

$$A_0 = 5K_1 \log K_1. \tag{3.49}$$

Each operation in (3.41) that tries to obtain one of the values of $\hat{X}(\omega_m)$ requires J complex multiplications and $J - 1$ complex additions. When M points need to be interpolated, the total number of real operations required by the interpolation is

$$A_1 = M[6J + 2(J - 1)] = M(8J - 2). \tag{3.50}$$

Because the interpolation step of the adjoint operator of the 1-D NUFFT uses the transpose matrix of that used in the 1-D NUFFT, the interpolation step of the adjoint operator of the 1-D NUFFT needs the same number of real multiplication operations as that of the 1-D NUFFT and the number of real addition operations needed by the interpolation step is approximately equal to that of the 1-D NUFFT. The fast computation of (3.44) can be realized by the IFFT of K_1 points, discarding all but the preceding K values. When IFFT is used to process (3.44), the sum of real operations required by IFFT [16, 18] is

$$A_2 = 5K_1 \log K_1. \tag{3.51}$$

When M nonuniform frequency points need to be interpolated, the total number of real operations required by the interpolation step of the adjoint operator of the 2-D NUFFT [18] should be

$$A_3 \approx A_1. \tag{3.52}$$

In general, to obtain the fastest computing speed, the interpolation coefficients are precomputed and stored in the memory of computers. The crucial problem for different fast algorithms of NUFFT and its adjoint operator is to determine the optimum interpolation coefficients and apply them in the computation.

The extensions of the 1-D NUFFT and its adjoint operator to 2-D case are conceptually straightforward. For the 2-D NUFFT, the 2-D FFT is employed and the interpolation coefficients are precomputed and stored for each desired frequency location [18]. Using the same analysis method in the 1-D NUFFT, for the 2-D NUFFT, the total number of real operations needed to generate M nonuniform frequency-domain samples with $K \times K$ equispaced signal samples is

$$\begin{aligned} A_4 &= M\left[6J^2 + 2(J^2 - 1)\right] + 5K_1K_2 \log(K_1K_2), \\ &= M\left(8J^2 - 2\right) + 5K_1K_2 \log(K_1K_2) \end{aligned} \tag{3.53}$$

where $K_1 = K_2 = \sigma K$.

As for the adjoint operator of the 2-D NUFFT, the sparse interpolation matrix used by the interpolation step (3.47) is the transpose of that used by the 2-D NUFFT and the fast computation of (3.48) can be realized by the 2-D IFFT. The number of real operations required by the interpolation of the adjoint operator of 2-D NUFFT with M nonuniform points should be $M\left[6J^2 + 2(J^2 - 1)\right]$ approximately. The number of real operations of the IFFT is the same as that of the FFT, $5K_1K_2\log(K_1K_2)$ shown in (3.53). Therefore, similar to the adjoint operator of 1-D NUFFT, the total number of real operations required to complete the adjoint operator of 2-D NUFFT with M nonuniform points [18] is

$$A_5 \approx A_4. \tag{3.54}$$

3.5.2 Beamforming with NUFFT

In the far-field condition, substituting (3.11) into (3.7), $B\left(r_0, l, \theta_{ap}, \theta_{eq}\right)$ can be rewritten as

$$B\left(r_0, l, \theta_{ap}, \theta_{eq}\right) = \sum_{m=1}^{M_0} w_m S_m(l) \exp\left[-i2\pi f_l\left(\frac{x_m \sin\theta_{ap} + y_m \sin\theta_{eq}}{c}\right)\right]. \tag{3.55}$$

Similar to the CZT beamforming, θ_{ai} and θ_{af} are the initial and final azimuth angles, and θ_{ei} and θ_{ef} represent the initial and final elevation angles. There are $\theta_{af} = -\theta_{ai}$ and $\theta_{ef} = -\theta_{ei}$ generally. Then the azimuth and elevation steering angles employed for the NUFFT beamforming method are also equispaced in the sine domain, given by

$$\sin\theta_{ap} = p\Delta s_a + \Delta s_a/2,\ \sin\theta_{eq} = p\Delta s_e + \Delta s_e/2 \quad \begin{pmatrix} -M_b/2 \le p \le M_b/2 - 1 \\ -N_b/2 \le q \le N_b/2 - 1 \end{pmatrix}. \tag{3.56}$$

where Δs_a and Δs_e are defined by (3.26) and (3.27) respectively.

Considering (3.56), (3.55) becomes

$$B(r_0, l, \theta_{ap}, \theta_{eq}) = \sum_{m=1}^{M_0} S'_m(l) \exp[i(p\omega_{am} + q\omega_{em})], \tag{3.57}$$

where

$$S'_m(l) = w'_m S_m(l), \tag{3.58}$$

$$w'_m = w_m \exp[-i2\pi f_l(x_m \Delta s_a + y_m \Delta s_e)/2c] \quad (3.59)$$

and

$$\omega_{am} = -2\pi f_l x_m \Delta s_a / c, \quad \omega_{em} = -2\pi f_l y_m \Delta s_e / c. \quad (3.60)$$

For the proposed beamforming method based on NUFFT, the maximum values of ω_{am} and ω_{em} should be lower than π. The expression of the limiting condition for the proposed beamforming method is given as

$$\begin{cases} \max\limits_m(\omega_{am}) = \frac{\pi f_H D}{c} \cdot \frac{\sin\theta_{af} - \sin\theta_{ai}}{M_b - 1} < \pi \\ \max\limits_m(\omega_{em}) = \frac{\pi f_H D}{c} \cdot \frac{\sin\theta_{ef} - \sin\theta_{ei}}{N_b - 1} < \pi \end{cases}, \quad (3.61)$$

where f_H is the highest frequency in the passband of the received signals.

If an equispaced 2-D array with a square aperture is employed and N_1 is the lateral size of the array, both x_m and y_m will be located between $-(N_1 - 1)\lambda_0/4$ and $(N_1 - 1)\lambda_0/4$ generally, where λ_0 is the wavelength of the carrier whose frequency is f_0. For the application of broadband 3-D imaging mentioned in the papers [6, 26], there are $0.5 f_0 < f_l < 1.5 f_0$, $\theta_{af} \leq 30°$ and $\theta_{ef} \leq 30°$. Moreover, to attain high imaging quality, the beams of imaging systems should be set as dense as possible. Thus, M_b and N_b are usually set to $2N_1$ [6, 26]. It is easy to see that the limiting condition (3.61) is satisfied. Using similar parameter configurations for square-aperture arbitrary arrays the same conclusion (i.e., $\omega_{am} \in [-\pi, \pi)$ and $\omega_{em} \in [-\pi, \pi)$) can be drawn.

For both equispaced and arbitrary 2-D arrays, (3.57) has the same form as (3.46), the expression of the adjoint 2-D NUDFT. Hence, the fast computation of (3.57) can be realized by the adjoint operator of the 2-D NUFFT.

According to the analysis of (3.53), the number of the real operations needed to compute (3.57) at a single frequency f_l for all beams, denoted by A_6, is

$$A_6 \approx M_0(8J^2 - 2) + 5M'_b N'_b \log_2(M'_b N'_b), \quad (3.62)$$

where due to the application of IFFT, M'_b should be the first powers of two, greater than or equal to σM_b, and N'_b should be the first powers of two, greater than or equal to σN_b.

The nonuniform frequency nodes ω_{am} and ω_{em} and the interpolation coefficients of the adjoint operator of the 2-D NUFFT are usually precomputed and stored for fast 3-D beamforming. In addition, the proposed method based on NUFFT has no special requirements for positions of the sensors, meaning that the sensors of 2-D arrays can be distributed arbitrarily. Therefore, the proposed method is suitable for arbitrary 2-D arrays.

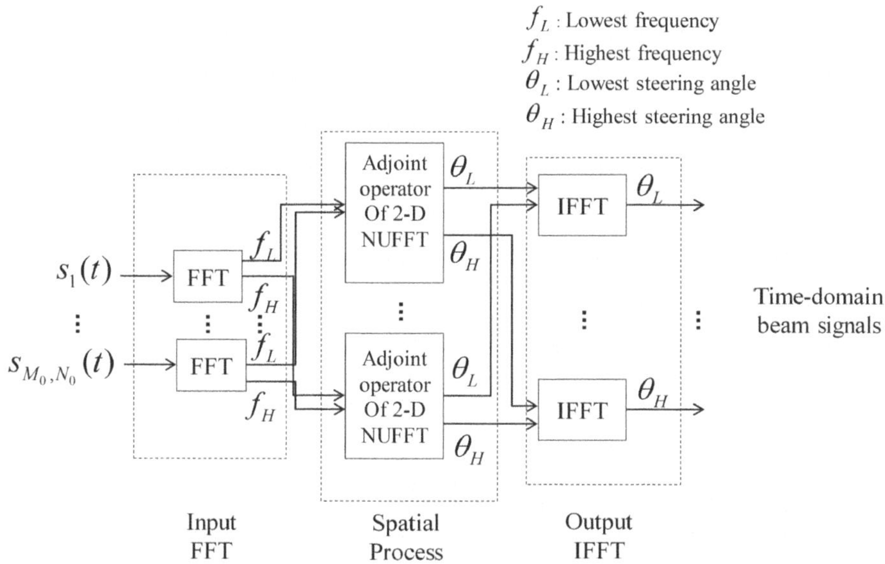

Fig. 3.6 Schematic of the NUFFT 3-D beamforming method

To sum up, the NUFFT 3-D beamforming method for underwater real-time 3-D acoustical imaging in both the far and near fields is realized by the following three steps:

(i) Input FFT: segment the received broadband signals into sequential blocks and convert the blocks into frequency-domain signals by using FFT.
(ii) Spatial processing: multiply the FFT coefficients of each sensor by the corresponding weights (near field: dynamic focusing) which are generated off line, then compute beam signals at every valid frequency point with the adjoint operator of the 2-D NUFFT.
(iii) Output IFFT: invert the beam signals in the frequency domain into the time-domain signals by using IFFT.

The procedure of the NUFFT 3-D beamforming [6] is shown in Fig. 3.6.

3.5.3 Accuracy Evaluation

Unlike the frequency-domain direct method of (3.7) and (3.6), which is an exact method, the proposed method is an approximated one because the realization of NUFFT is an approximated method. In this sub-chapter, the beamforming errors caused by NUFFT are evaluated and proven to be negligible in the application of 3-D underwater acoustical imaging in both the far and near fields.

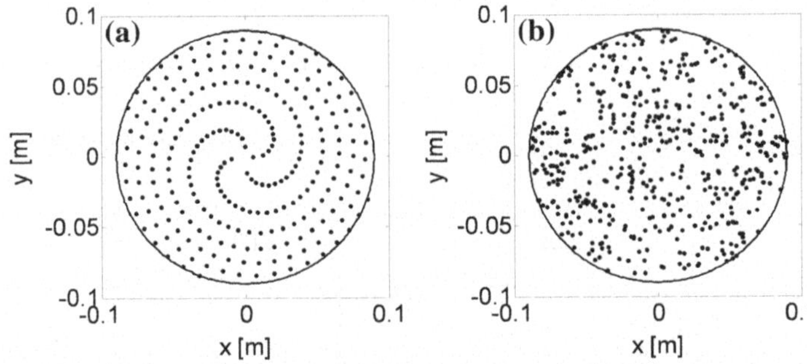

Fig. 3.7 **a** A spiral 2-D array with 256 elements. **b** A random 2-D array with 512 elements

In addition, although the CZT 3-D beamforming [3–5] cannot be directly applied to arbitrary arrays with off-the-grid elements, it is possible to be applied by approximating the array element positions with their closest on-the-grid positions. The accuracy loss of the approximated CZT is also discussed and compared with that of NUFFT method to verify the advantage of NUFFT.

The computational accuracy of NUFFT is determined by the two parameters: σ and J, as is mentioned above. To evaluate the influence of the two parameters on the accuracy of beamforming, NUFFT is realized based on the toolbox given by Fessler [27], where σ and J can be set to different values.

Two arbitrary arrays: a spiral array and a random array, as depicted in Fig. 3.7a and b are employed for computing the errors caused by NUFFT. The diameter of the two arrays is $60\lambda_0$, where λ_0 is the wavelength of the carrier frequency f_0. For a fully sampled uniform 2-D circular array with the diameter $60\lambda_0$, the total number of elements will be 11,305. The spiral 2-D array is composed of 256 omnidirectional punctiform elements, whose divergence angle, the key parameter of the spiral array, is 92.05°, as given in [28]. The random 2-D array with uniform distribution [29, 30] is composed of 512 elements. The transmitter located in the center of the array is omnidirectional. A 3-D scene with two reflecting scatterers is considered, the positions of which are located at $(\rho, \theta_a, \theta_e) = (20\,\text{m}, -15°, -15°)$ and $(20\,\text{m}, 0°, 0°)$, as described by the geometry of Fig. 3.1a, where ρ denotes the distance from the center of the 2-D array to the scatterer. The scatterers are all taken as points and the medium is supposed to be isotropic and ideal without any frequency absorption. The transmitting signal is a Gaussian-envelope pulse with central frequency f_0 being 500 kHz and the 3 dB bandwidth being 180 kHz. This situation is obviously corresponding to the far-field condition. The received signals of the sensors are simulated by using the method adopted by Palmese and Trucco [31]. The details about the simulation technique will be given in Chap. 5.

The relative error $\varepsilon_{\max}$ of the whole 3-D image obtained by using the NUFFT 3-D beamforming method can be computed [18] by

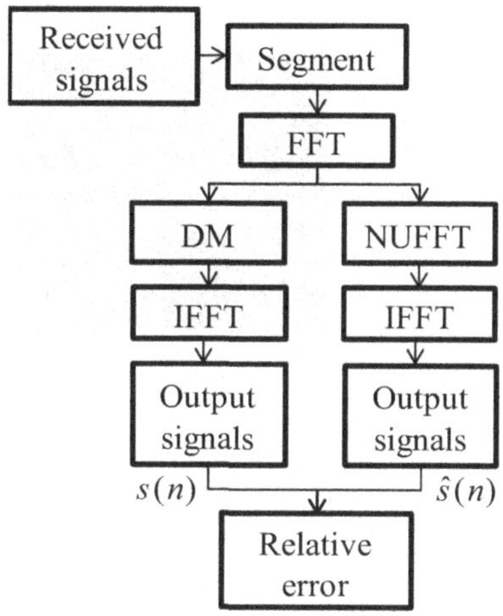

Fig. 3.8 Procedure of computing the relative errors of the whole 3-D image scene. (DM; the frequency-domain direct method)

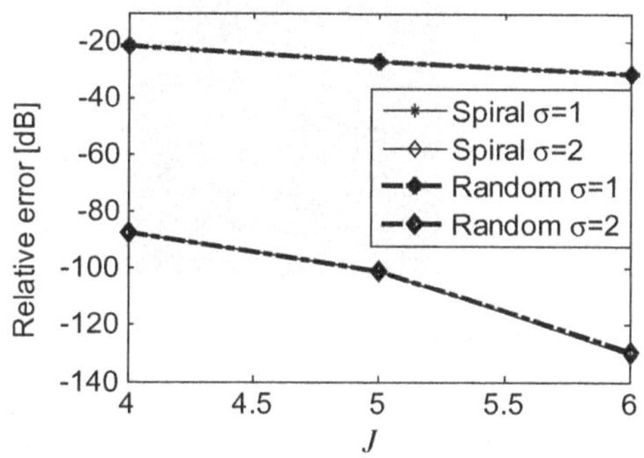

Fig. 3.9 Relative errors of the whole 3-D image with different values of σ and J

$$\varepsilon_{\max} = 20\log_{10}\left(\frac{\max_{n,m}\left(\left|\hat{s}_m(n) - s_m(n)\right|\right)}{\max_{n,m}(|s_m(n)|)}\right), \tag{3.63}$$

where $\hat{s}_m(n)$ and $s_m(n)$ represent the mth beam signal obtained by the NUFFT 3-D beamforming method and the frequency-domain direct method, respectively, and "$\max_{n,m}(\ )$" represents the maximum of the expression in the bracket for all the values of n and m. The procedure that computes the relative errors is depicted in Fig. 3.8. The relative errors for different values of σ and J are computed and depicted in Fig. 3.9. From Fig. 3.9, it can be found that for both arrays, the relative errors are all above -32 dB when $\sigma = 1$ and the relative errors will become less than or equal to -87 dB when $\sigma = 2$, with J being assigned different values between 4 and 6.

In order to reflect the accuracy of the proposed method intuitively, Fig. 3.10 (for the spiral array) and 3.11 (for the random array) show the 2-D slices at the focusing distance of the whole 3-D images obtained by the frequency-domain direct method and the NUFFT 3-D beamforming ($J = 4$ and $\sigma = 2$). It can be found that there

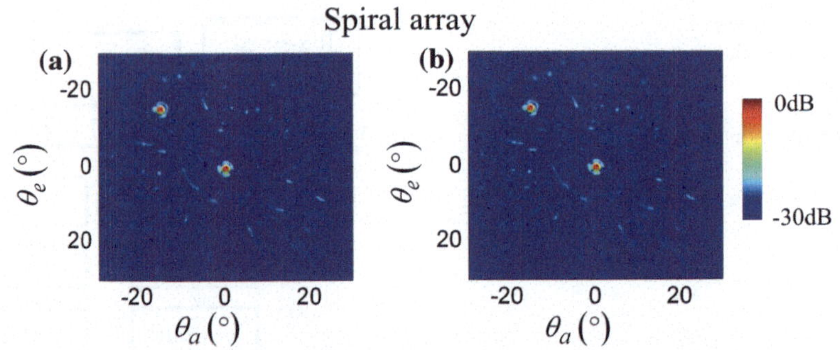

Fig. 3.10 The 2-D slices at the focusing distance of the 3-D images obtained by the frequency-domain direct method and the NUFFT 3-D beamforming ($J = 4$ and $\sigma = 2$) for the spiral array. **a** for the frequency-domain direct method; **b** for the NUFFT 3-D beamforming

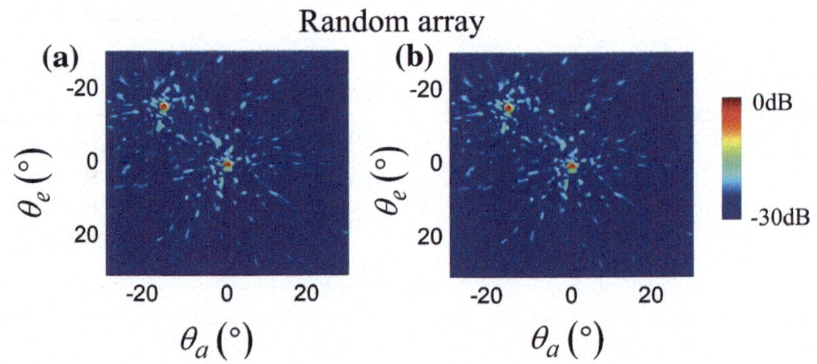

Fig. 3.11 The 2-D slices at the focusing distance of the 3-D images obtained by the frequency-domain direct method and the NUFFT 3-D beamforming ($J = 4$ and $\sigma = 2$) for the random array. **a** for the frequency-domain direct method; **b** for NUFFT 3-D beamforming

is nearly no difference between the 2-D slices obtained by the two methods for the different arrays. The relative errors of the side lobe level of the 2-D slices are all lower than −80 dB. The relative errors of the main lobe width are also lower than −80 dB. The output beam signals in the direction (−15°, −15°) and (15°, 15°) computed by the frequency-domain direct method and the NUFFT 3-D beamforming ($J = 4$ and $\sigma = 2$) are given in Fig. 3.12. It is found that the two beam signals almost overlap with each other. The figures demonstrate that the proposed method based on NUFFT has a good accuracy in the far field.

To verify the high accuracy of the NUFFT 3-D beamforming further, the accuracy loss of the approximated CZT beamforming is also discussed and compared with that of the NUFFT method. The approximated CZT beamforming is realized by approximating the element positions of the spiral and random arrays with their closest on-the-grid positions and then applying the CZT [11, 18]. The output time-domain beam signals of the approximated CZT method applied to the two arrays are shown

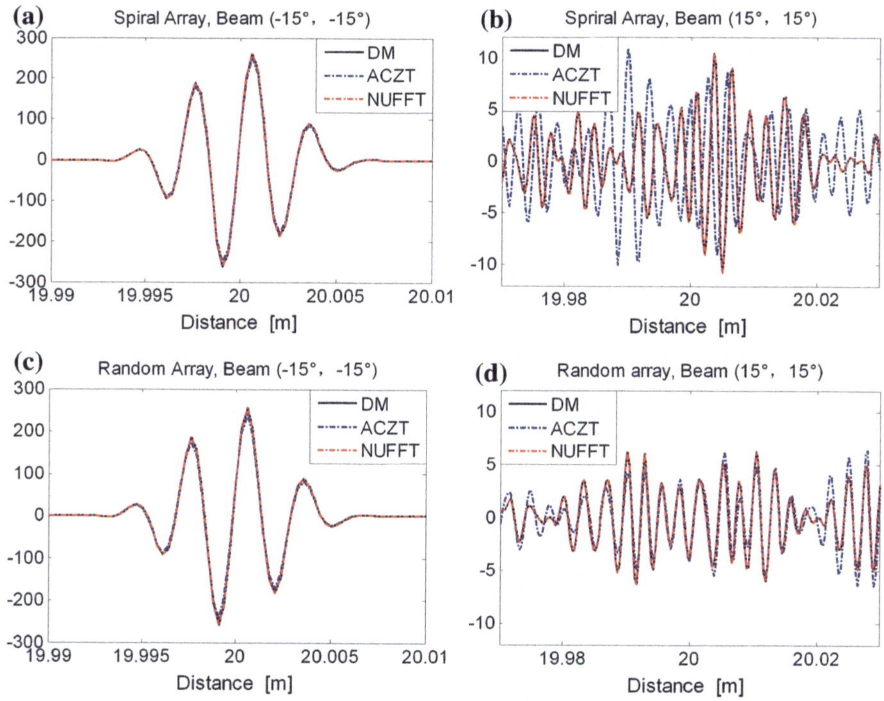

Fig. 3.12 Output beam signals of the frequency-domain direct method (DM), the approximated CZT method (ACZT) and the NUFFT 3-D beamforming method ($J = 4$ and $\sigma = 2$) for the spiral and random arrays. **a** Beam $(-15°, -15°)$ for the spiral array; **b** Beam $(15°, 15°)$ for the spiral array; **c** Beam $(-15°, -15°)$ for the random array; **d** Beam $(15°, 15°)$ for the random array

Table 3.1 The relative errors of the whole 3-D image caused by the proposed method based on NUFFT ($J = 4$ and $\sigma = 2$) and the approximated CZT method (ACZT) applied to the spiral and random arrays

Array pattern	Relative error for ACZT (dB)	Relative error for NUFFT (dB)
Spiral	−21.7145	−87.3355
Random	−23.6289	−87.2251

in Fig. 3.12. It can be seen that in the direction (15°, 15°) where exist no scatterers, the beam signals of the approximated CZT method deviate obviously from that of the frequency-domain direct method, while the beam signals of the proposed method still overlap with that of the frequency-domain direct method. The relative errors of the whole 3D image caused by the approximated CZT and the proposed methods are given in Table 3.1. It can be found that the relative errors caused by the approximated CZT method are much higher than those caused by the NUFFT 3-D beamforming (Fig. 3.11).

Table 3.2 The relative errors of the whole 3-D image in the near-field caused by the proposed method based on NUFFT ($J = 4$ and $\sigma = 2$) and the Fresnel approximation direct method (FADM) applied to the spiral and random arrays

Array pattern	Relative error for FADM (dB)	Relative error for NUFFT (dB)
Spiral	−30.2154	−30.2150
Random	−25.2324	−25.2320

The above analyses are under the far-field condition, which might not be enough to verify the accuracy of the proposed method. In the following, the relative errors are discussed in the near field for the spiral and random arrays. In the near field, the errors of the proposed method are caused by NUFFT and the Fresnel approximation. To evaluate the errors caused by the Fresnel approximation, the direct computation of (3.15) is conducted, which is called as the Fresnel approximation direction method (FADM). The direction computation of (3.7), namely the frequency-domain direct method which is accurate both in the near and far fields, is applied to provide reference results.

A near-field 3-D scene with two reflecting scatters is considered. The two near-field scatterers are located at $(\rho, \theta_a, \theta_e) = (1\,\text{m}, -5°, -5°)$ and $(1\,\text{m}, 0°, 0°)$ respectively. Figure 3.13 shows that in the direction $(10°, 10°)$ where exist no scatters, the beam signals of the proposed method based on NUFFT ($J = 4$ and $\sigma = 2$) deviate obviously from those of the frequency-domain direct method but almost overlap with those of the FADM. Table 3.2 illustrates that the relative errors caused by the FADM and the proposed method based on NUFFT ($J = 4$ and $\sigma = 2$) are nearly the same for both the spiral and random arrays. It can be concluded from Fig. 3.13 and Table 3.2 that the error caused by NUFFT ($J = 4$ and $\sigma = 2$) are negligible in the near field, compared with that caused by the Fresnel approximation.

On the basis of all the analyses, it comes to a conclusion that $\sigma = 2$ can be adopted to realize the NUFFT 3-D beamforming, and the relative errors are so small that they can be neglected in the application of underwater real-time 3-D acoustical imaging.

3.6 Compuational Load for Direct Method, CZT and NUFFT Beamforming

We evaluates the minimum number of on-line real operations needed to generate a 3-D image in two situations: one with an equispaced 2-D array and the other with an arbitrary 2-D array. Four beamforming methods are considered in the evaluation. They are: (1) the time-domain delay-and-sum method; (2) the frequency-domain direct method; (3) the CZT method; (4) the NUFFT method.

The number of on-line real operations includes both real additions and real multiplications. Any operation that can be performed off line (before the start of the 3-D

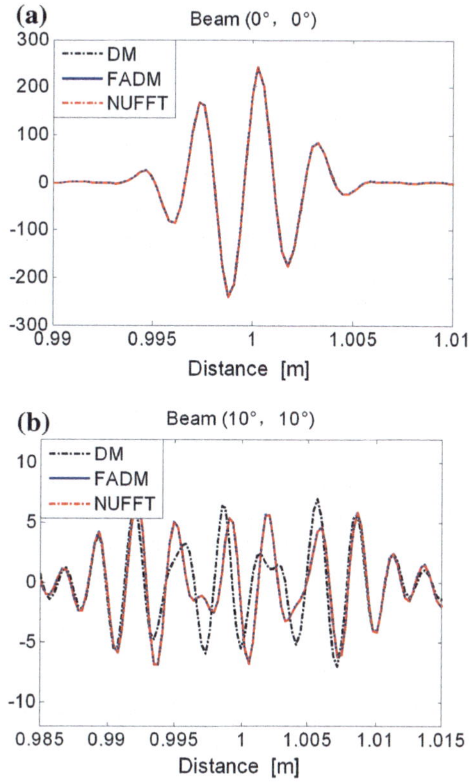

Fig. 3.13 Output near-field beam signals of the frequency-domain direct method (DM), the Fresnel approximation direct method (FADM) and the NUFFT 3-D beamforming ($J = 4$ and $\sigma = 2$) for the spiral array. **a** Beam (0°, 0°); **b** Beam (10°, 10°)

image generation) is disregarded [3]. The main parameters for the evaluation are described as follows:

- the number of sensors of the equispaced 2-D array: $M_0 = N_1 \times N_1$;
- the number of sensors of the arbitrary 2-D array: M_1;
- square set of beams: $N_b \times N_b$.

3.6.1 Equispaced 2-D Arrays

When an equispaced 2-D array is employed, the CZT beamforming proposed by Palmese and Trucco is valid [3, 4]. The computational loads needed by the four beamforming methods are analyzed respectively.

For the time-domain delay-and-sum beamforming, an finite-impulse response (FIR) interpolation filter with H stages [9] is adopted for the signal oversampling before beamforming. The rate of real operations [3, 9] needed to generate N_b^2 beams is

$$TD_1 = \left[N_1^2(H + N_b^2) + N_1^2(H + 1)\right] f_s. \tag{3.64}$$

The computation of delays is considered an offline operation.

For the three frequency-domain beamforming methods aforementioned, they all consist of three steps: input FFT, spatial processing and output IFFT. The crucial factor deciding the number of on-line real operations of the three frequency-domain beamforming methods is their algorithms of the spatial processing.

The first step of the three frequency-domain beamforming methods is to segment the time-domain received signals and convert them into the frequency-domain coefficients through FFT. The number of on-line real operations needed to process a L-length real sequence by FFT [4, 16] is

$$F_{FFT} = 2.5L \log \frac{L}{2} + 7L. \tag{3.65}$$

In the second step, the frequency-domain direct method is realized by a complex dot product described in Sect. 2. As shown in [3, 4], the spatial processing of the frequency-domain direct method that generates N_b^2 beams at a single frequency f_l requires the following number of on-line real operations:

$$F_{Dsp} = \left(8N_1^2 - 2\right)N_b^2. \tag{3.66}$$

To generate N_b^2 beams, the spatial processing of the CZT beamforming method at a single frequency f_l requires the following number of on-line real operations [4]:

$$F_{Csp} = 6\left[N_1^2 + N_b^2 + L_1^2\right] + 20L_1^2 \log_2(L_1), \tag{3.67}$$

where L_1, the size of FFT, should be a power of two, and $L_1 \geq N_1 + N_b - 1$ due to the use of FFT and IFFT.

The proposed beamforming method based on NUFFT is to transform the spatial processing into the adjoint operator of 2-D NUFFT. The spatial processing of the method includes the following steps:

(1) compute w'_m, w''_m and the interpolation coefficients (off line);
(2) multiply $S_m(l)$ by w'_m or w''_m to obtain $S'_m(l)$ or $S''_m(l)$, and then interpolate $S'_m(l)$ or $S''_m(l)$(on line) based on the adjoint operator of 2-D NUFFT;
(3) use IFFT to process $\tilde{S}'_m(l)$ or $\tilde{S}''_m(l)$ obtained by interpolating (on line).

For multiplying $S_m(l)$ by w'_m or w''_m to obtain $S'_m(l)$ or $S''_m(l)$, N_1^2 complex multiplication operations are needed, which are equal to $6N_1^2$ real operations. According to (3.53), to generate N_b^2 beams, the number of on-line real operations required by the spatial processing of the proposed method based on NUFFT at a single frequency f_l is:

$$F_{Nsp} \approx 6N_1^2 + N_1^2\left(8J^2 - 2\right) + 5N_b'^2 \log_2(N_b'^2), \tag{3.68}$$

where N'_b should be the first power of two, greater than or equal to σN_b due to the application of IFFT.

The last step for all the three frequency-domain methods: direct, CZT and NUFFT, is to obtain time samples of each beam signal by applying IFFT. To further reduce the computational load, the technique of generating cubic resolution cells proposed by Palmese and Trucco is adopted [4], which is briefly described as follows.

Owing to the fact that the angular resolution of the beamforming system is constant, the lateral resolution worsens with the increase of the distance. But the range resolution, depending on the pulse bandwidth, does not vary with the increase of the distance. Subsequently, the technique of generating cubic resolution cells [4] makes it possible to avoid unnecessary operations by reducing the number of frequency bins processed (i.e., the bandwidth considered) with the increase of the distance. Assuming that Q is the bandwidth of the continuous-wave (CW) pulse, the range resolution is $c/(2Q)$ [4]. Considering that λ is the wavelength of the carrier and d is the intersensor spacing, the lateral resolution of the equispaced array [4] which is achieved by weighting is [4]

$$R_0 = 0.88\lambda r/D_1, \tag{3.69}$$

where r is the given distance, and D_1 is the side size of the 2-D square arrays employed in this book. In particular, for the equispaced 2-D array,

$$D_1 = N_1 d. \tag{3.70}$$

To make the range resolution and lateral resolution equal, the bandwidth [4] should be

$$Q = cNd/1.76\lambda r. \tag{3.71}$$

Then the number of frequency bins needed to achieve a range resolution not worse than the lateral resolution [4] is the following:

$$\zeta = \text{ceil}\left(cLN_1 d/1.76 f_s \lambda r^*\right), \tag{3.72}$$

where $\text{ceil}(x)$ represents the first integer greater than x, and r^* denotes the distance at which the signal segment starts.

Finally, the frequency bins are shifted around $f = 0$ to obtain the equivalent lowpass spectrum. The envelop used to display a 3-D image is extracted with low sampling frequency. The number of on-line real operations of IFFT is:

$$F_{IFFT} = 5\zeta_1 \log_2(\zeta_1), \tag{3.73}$$

where ζ_1 is the first power of two which is greater than or equal to ζ.

For practical systems, the equispaced 2-D array can be thinned to form a sparse array with the active fraction T of the sensors in order to reduce the hardware cost. The computational load of the time-domain delay-and-sum beamforming is certainly mitigated with the reduction of active sensors. As for the three frequency-domain methods, the number of on-line real operations required in the input FFT step can also be greatly decreased with the reduction of active sensors. However, the spatial processing is different among the three frequency-domain methods. For the spatial processing of the frequency-domain direct method and the proposed method based on NUFFT, the number of real operations can be decreased through the reduction of active sensors. For the CZT method, the number will remain unchanged since the spatial processing of the CZT method utilizes FFT and IFFT to realize the fast computation of convolution [3, 4] and the data of these inactive sensors which are set to zero are still employed by the spatial processing of the CZT method.

3.6.2 Arbitrary 2-D Arrays

Arbitrary 2-D arrays usually can be referred to 2-D arrays composing of arbitrarily distributed sensors and are a solution to help further cut the number of sensors and achieve better directivity than equispaced 2-D arrays with same apertures. In this case, the CZT beamforming is not available and $N_1 \times N_1$ is not applicable for describing the number of sensors. Thus, the number of sensors of the arbitrary 2-D array is assumed to be M_1.

Based on (3.64), for the arbitrary 2-D array, the number of real operations [4, 10] needed by the time-domain delay-and-sum method to generate N_b^2 beams is

$$TD_1 = \left[M_1(H + N_b^2) + M_1(H + 1)\right] f_s. \tag{3.74}$$

With regard to the arbitrary 2-D array, according to (3.66), the number of on-line real operations of the spatial processing to generate N_b^2 beams at a single frequency f_l for the DM is

$$F'_{Dsp} = (8M_1 - 2)N_b^2. \tag{3.75}$$

According to (3.68), the number of on-line real operations required by the spatial processing of the proposed method based on NUFFT at a single frequency f_l is

$$F'_{Nsp} \approx 6M_1 + M_1\left(8J^2 - 2\right) + 5N_b'^2 \log_2(N_b'^2). \tag{3.76}$$

3.6.3 Comparison

Giving the array size and the size of beam set, the total number of on-line real operations of different beamforming methods required by generating a 3-D underwater image can be computed by the equations introduced in Sect. 3.6.3. The main parameters for the computation of underwater real-time 3-D imaging are as follows:

- sound speed: $c = 1500$ m/s;
- carrier frequency: 500 kHz;
- signal bandwidth: $B = 500$ kHz;
- sampling frequency: $f_s = 1.5$ MHz;
- effective range: 1–100 m (including both the near and far fields);
- intersensor spacing of the equispaced 2-D array: $d = 1.5$ mm;
- length of signal segments: $L = 1024, 2048, 4096$;
- stages of the interpolation filter: $H = 100$;

The influence of different values of L on the computational load has been discussed in the papers of Palmese and Trucco [3, 4]. Based on the analysis of their papers, in the near-field condition (Fresnel zone), the value of L in this paper is either 1024 or 2048, dynamically determined by the DOF [4]. In the far-field condition, L is fixed to 4096. For all the frequency-domain methods, the technique of generating cubic resolution cells [4] is applied to reduce the number of operations further.

First, consider an equispaced 2-D array for which the CZT beamforming method is feasible. As mentioned previously, the difference among the three frequency-domain methods is their algorithms of spatial processing. In the algorithms of spatial processing, the ratio N_b/N_1 and the length L of a data segment are assumed to be 2 and 4096, respectively. As the lateral size of the equispaced array N_1 changes, the number of on-line real operations required by the spatial processing algorithms of the three frequency-domain methods at a valid frequency bin will changes, which is shown in Fig. 3.14. Figure 3.14 illustrates that the number of real operations needed by the spatial processing of the direct method is at least one order of magnitude higher than that needed by the spatial processing of the CZT method and the NUFFT beamforming method ($J = 4$ and $\sigma = 2$) for $N_1 \geq 74$. For the spatial processing of the NUFFT beamformig method, a jump occurs between $N_1 = 64$ and $N_1 = 65$. The reason is that the size of IFFT N_b' used in the adjoint operator of NUFFT jumps from 256 to 512. For the spatial processing of the CZT method, a jump occurs between $N_1 = 85$ and $N_1 = 86$, because the size of FFT L_1 is changed from 256 to 512. In most cases, the number of operations for the spatial processing of the proposed method based on NUFFT is lower than that required by the CZT spatial processing except for $N_1 = 65, \ldots, 86$.

The total number of on-line real operations required by the four beamforming methods (i.e., NUFFT, CZT, direct method and delay-and-sum) for generating a 3-D image is given in Table 3.3. The results show that when the sensors of the array are fully activated, i.e., the active fraction $T = 100\%$, the total number of on-line real operations of the NUFFT method is obviously lower than that of the frequency-domain direct method and the time-domain delay-and-sum method, and a little bit

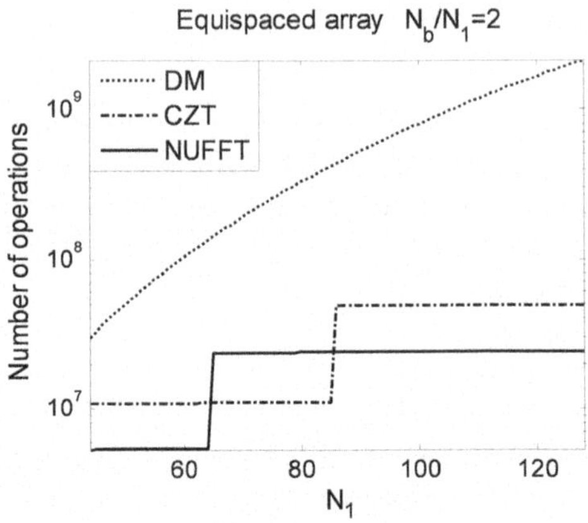

Fig. 3.14 The number of operations of different spatial processing algorithms versus the lateral size N_1 ($J = 4$ and $\sigma = 2$ for NUFFT). (DM: the frequency-domain direct method)

Table 3.3 Number of on-line real operations to create a 3-D image with the equispaced arrays ($J = 4$, $\sigma = 2$ for NUFFT)

	NUFFT	CZT	DM	D&S
$N_1 = 64,\ N_b = 128,\ T = 100\%$	2.47×10^{10}	2.61×10^{10}	1.29×10^{11}	1.05×10^{13}
$N_1 = 128,\ N_b = 256,\ T = 100\%$	1.20×10^{11}	1.34×10^{11}	4.53×10^{12}	2.15×10^{14}
$N_1 = 128,\ N_b = 256,\ T = 12.5\%$	2.93×10^{10}	4.24×10^{10}	5.70×10^{11}	2.69×10^{13}

(DM: the frequency-domain direct method; D&S: the time-domain delay-and-sum beamforming)

lower than that of the CZT method both at $N_1 = 64$ and $N_1 = 128$. Compared with the direct method and the delay-and-sum method, the NUFFT method shows an approximately 40-fold and 1800-fold gains at $N_1 = 128$, respectively. When the active fraction $T = 12.5\%$ at $N_1 = 128$, the NUFFT method achieves an 1.6-fold gain compared with the CZT method, and 20-fold and 920-fold gains compared with the direct method and the delay-and-sum method.

Second, consider an arbitrary array with nonuniform spacing and evaluate the computational load. Since the CZT beamforming method is not feasible, the comparison is conducted among the other three beamforming methods (i.e., NUFFT, direct method and delay-and-sum). For $M_1 = 8000$, the nonuniform array is assumed to have the same aperture as the square equispaced array whose lateral size is $N_1 = 128$. For $M_1 = 2000$, the nonuniform array has the same aperture as the square equispaced array whose lateral size is $N_1 = 64$. Figure 3.15 shows the computational weight of the three steps (i.e., input FFT, spatial processing, output IFFT) to compute a segment with $L = 4096$, needed by the direct method and the NUFFT method when $J = 4$ and $\sigma = 2$. It is easy to see that the computational loads of the direct method and the

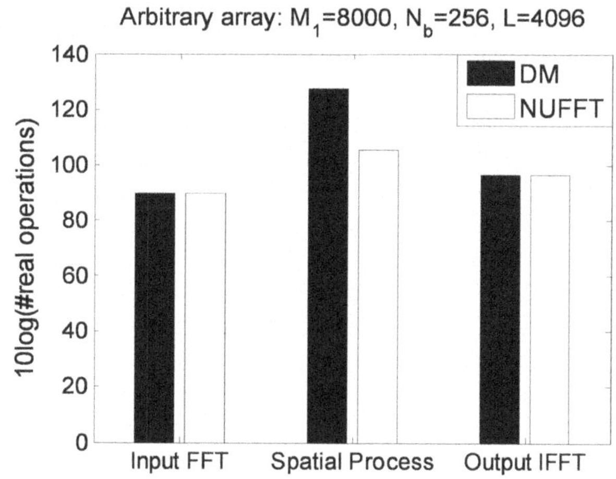

Fig. 3.15 Computational load of the three steps needed to compute a segment with $L = 4096$ for the frequency-domain direct method (DM) and the NUFFT beamforming method ($J = 4$ and $\sigma = 2$)

Table 3.4 Number of on-line real operations to create a 3-D image with the arbitrary array employed ($J = 4$, $\sigma = 2$ for NUFFT)

	NUFFT	DM	D&S
$M_1 = 2000$, $N_b = 128$	1.62×10^{10}	1.49×10^{11}	6.63×10^{13}
$M_1 = 8000$, $N_b = 256$	8.20×10^{10}	4.26×10^{12}	1.05×10^{14}

(DM: the frequency-domain direct method; D&S: the time-domain delay-and-sum beamforming)

NUFFT method in the first and third steps are equal, but for the spatial processing the number of real operations of the direct method is about 170-fold higher than that of the NUFFT method.

Table 3.4 gives the total number of on-line real operations required by the three beamforming methods to generate a 3-D image. It is shown that compared with the frequency-domain direct method and the time-domain delay-and-sum method, the NUFFT method has approximately 52-fold and 1280-fold computational gains at $M_1 = 8000$. The larger the number of sensors is, the higher the computational gain will be.

3.6.4 Summary

Unlike the CZT beamforming method which is only applicable for equispaced 2-D arrays and sparse arrays thinned from equispaced 2-D arrays, the NUFFT beamforming method is feasible for arbitrary arrays. The number of on-line real operations required by the NUFFT beamforming method is dramatically reduced, compared with that of the time-domain delay-and-sum beamforming and the frequency-domain direct method both for equispaced and arbitrary 2-D arrays. The gain factor of real operations obtained by using the NUFFT beamforming method is one or two orders of magnitude greater than that by using the frequency-domain direct method. When compared with the time-domain delay-and-sum beamforming method, the gain factor can reach three orders of magnitude. As for equispaced 2-D arrays, the NUFFT beamforming method still performs better than the CZT beamforming method in terms of computational load in most cases.

Although the NUFFT beamforming is an approximated method, with proper parameter configurations, the relative error of the output beam signal caused by NUFFT is usually lower than or equal to -87 dB, which is so small that it can be neglected for underwater real-time 3-D acoustical imaging.

References

1. A. Trucco, Thinning and weighting of large planar arrays by simulated annealing. IEEE Trans. Ultrason. Ferroelectr. Freq. Control **46**(2), 347–355 (1999)
2. V. Murino, A. Trucco, Underwater 3D imaging by FFT dynamic focused beamforming, in *Proceedings of 1st IEEE International Conference Image Processing*, vol. 1 (Austin, TX, 1994), pp. 890–894
3. M. Palmese, A. Trucco, Three-dimensional acoustic imaging by chirp zeta transform digital beamforming. IEEE Trans. Instrum. Meas. **58**(7), 2080–2086 (2009)
4. M. Palmese, A. Trucco, An efficient digital CZT beamforming design for near-field 3-D sonar imaging. IEEE J. Ocean. Eng. **35**(3), 584–594 (2010)
5. M. Palmese, A. Trucco, Pruned chirp zeta transform beamforming for 3-D imaging with sparse planar arrays. IEEE J. Ocean. Eng. **39**(2), 206–211 (2014)
6. C. Chi, Z. Li, Q. Li, Fast broadband beamforming using nonuniform fast Fourier transform for underwater real-time 3-D acoustical imaging. IEEE J. Ocean. Eng. **41**(2), 249–261 (2016)
7. D. Zhao, X. Liu, W. Chen, Y. Chen, Optimized designed for sparse cross arrays in both near-field and far-field. IEEE J. Ocean. Eng. Early Access **99**, 1–13 (2018)
8. V. Murino, A. Trucco, Three-dimensional image generation and processing in underwater acoustic vision. Proc. IEEE **88**(12), 1903–1948 (2000)
9. R.A. Mucci, A comparison of efficient beamforming algorithms. IEEE Trans. Acoust. Speech Signal Process. **32**(3), 548–558 (1984)
10. R.O. Nielsen, *Sonar Signal Processing, Chap. 2* (Artech House, Boston, 1991)
11. B. Maranda, Efficient digital beamforming in the frequency domain. J. Acoust. Soc. Am. **86**(5), 1813–1819 (1989)
12. A. C. Dhanantwari, S. Stergiopoulos, L. Song, C. Parodi, F. Bertora, P. Pellegretti, A. Questa, An efficient 3D beamformer implementation for real-time 4D ultrasound systems deploying array probes, in *proceedings of IEEE ultrasonics symposium* (2004), pp. 1421–1424
13. P. Chen, X. Tian, Y. Chen, Optimization of the digital near-field beamforming for underwater 3-D sonar imaging system. IEEE Trans. Instrum. Meas. **59**(2), 415–424 (2010)
14. A. Trucco, A least-squares approximation for the delays used in focused beamforming. J. Acoust. Soc. Am. **104**(1), 171–175 (1998)
15. L.J. Ziomek, Three necessary conditions for the validity of the Fresnel phase approximation for the near-field beam pattern of an aperture. IEEE J. Ocean. Eng. **18**(1), 73–75 (1993)
16. V.A. Oppenheim, R.W. Schafer, J.R. Buck, *Discrete-Time Signal Processing, Chap. 9* (Prentice-hall, Englewood Cliffs, 1989)

17. A. Dutt, V. Rokhlin, Fast Fourier transforms for nonequispaced data II. Appl. Comput. Harmon. Anal. **2**, 85–100 (1995)
18. J.A. Fessler, B.P. Sutton, Nonuniform fast Fourier transforms using min-max interpolation. IEEE Trans. Signal Process. **51**(2), 560–574 (2003)
19. L. Sha, H. Guo, W. Song, An improved gridding method for spiral MRI using nonuniform fast Fourier transform. J. Magn. Reson. **162**(2), 250–258 (2003)
20. D. Potts and G. Steidl, New Fourier reconstruction algorithms for computerized tomography, in *International Symposium on Optical Science and Technology* (2000), pp. 13–23
21. L. Michael and A. M. Zoubir, Fast wideband near-field imaging using the non-equispaced FFT with application to through-wall radar, in 19th European Signal Processing Conference (EUSIPCO), (Barcelona, 2011), pp. 1708–1712
22. P. Kruizinga, F. Mastik, N. de Jong, F.W. van der Steen, G. van Soest, Plane-wave ultrasound beamforming using a nonuniform fast Fourier transform. IEEE Trans. Ultrason. Ferroelectr. Freq. Control **59**(12), 2684–2690 (2012)
23. K. Yang, Z. Zhao, Q.H. Liu, Fast pencil beam pattern synthesis of large unequally spaced antenna arrays. IEEE Trans. Antennas Propag. **61**(2), 627–633 (2013)
24. M. Haltmeier, O. Scherzer, G. Zangerl, A reconstruction algorithm for photoacoustic imaging based on the nonuniform FFT. IEEE Trans. Med. Imaging **28**(11), 1727–1735 (2009)
25. S. Kunis, D. Potts, Time and memory requirements of the nonequispaced FFT. preprints 2006-01, TU-Chemnitz (2006)
26. A. Trucco, M. Palmese, S. Repetto, Devising an affordable sonar system for underwater 3-D vision. IEEE Trans. Instrum. Meas. **57**(10), 2348–2354 (2008)
27. J. A. Fessler, Image reconstruction toolbox (July, 2014). http://www.eecs.umich.edu/~fessler/code/index.html
28. O. Martínez-Graullera, C.J. Martín, G. Godoy, L.G. Ullate, 2D array design based on Fermat spiral for ultrasound imaging. Ultrasonics **50**, 280–289 (2010)
29. W.J. Hendricks, The totally random versus the bin approach for random arrays. IEEE Trans. Antennas Propag. **39**(12), 1757–1762 (1991)
30. R.E. Davidsen, J.A. Jensen, S.W. Smith, Two-dimensional random arrays for real-time volumetric imaging. Ultrason. Imaging **16**, 143–163 (1994)
31. M. Palmese, A. Trucco, Acoustic imaging of underwater embedded objects: signal simulation for three-dimensional sonar instrumentation. IEEE Trans. Instrum. Meas. **55**(4), 1339–1347 (2006)

Chapter 4
Design of Underwater Large Sparse 2-D Arrays

Abstract A large two-dimensional (2-D) array is mandatory for underwater real-time three-dimensional (3-D) acoustical imaging. To reduce the hardware cost, we need to design sparse large 2-D arrays. In different bandwidths, the design methods should be different. In this chapter, the design methods are presented in narrowband, wideband, and ultrawideband (UWB) respectively. Currently, the most popular narrowband method of designing large 2-D arrays for underwater real-time 3-D imaging is based on simulated annealing. Unfortunately, designing wideband large 2-D arrays is generally limited by the high computational load of computing wideband beam pattern. A fast method of computing wideband beam pattern is introduced. Then, a wideband design example based on the fast method of computing wideband beam pattern is shown. This chapter also reveals that UWB ultrasparse 2-D arrays are promising for obtaining ultralow hardware cost with keeping the same imaging quality. Ultralarge UWB 2-D arrays can help achieve the high angular resolution 0.1° with low cost.

Keywords Array design · Large 2-D arrays · Narrowband · Sparse array · Ultrawideband array · Ultrasparse array · Wideband beam pattern · Wideband array

4.1 Concept of Designing Large 2-D Arrays for Underwater 3-D Imaging

The techniques of sparse array designs have been researched for more than 60 years [1–3]. The aim of designing sparse arrays is to minimize the number of sensors under the requirements of directivity and signal-to-noise ratio (SNR). As analyzed in Chap. 1, the angular resolution of conventional underwater real-time 3-D acoustical imaging systems is around 1°. The size of 2-D arrays should be around $50\lambda_0 \times 50\lambda_0$, where λ_0 is the wavelength of the carrier frequency f_0. To obtain a higher angular resolution, a larger size is needed. Thus, it is emphasized that only the methods of designing large 2-D arrays for underwater real-time 3-D acoustical imaging are discussed in this chapter. This chapter does not include those design methods for

C. Chi, *Underwater Real-Time 3D Acoustical Imaging*, Signals and Communication Technology, https://doi.org/10.1007/978-981-13-3744-4_4

designing one-dimensional or smalls 2-D arrays. In addition, this chapter employs the coordinate geometry and notation, which are similar to that used in Chap. 3.

Reference [2] classifies the design methods of sparse arrays into three approaches: (1) optimize the positions and the weights of sensors simultaneously; (2) design on a priori fixed pattern of unequally spaced elements (e.g. random array), and optimize the weights of sensors; (3) design the positions of sensors without weighting. The narrowband design methods in this chapter follow the first approach. The wideband and UWB design methods adopt the third approach currently.

4.2 Narrowband 2-D Array

4.2.1 Definition of Narrowband Beam Pattern

The coordinate geometry shown in Fig. 4.1 is usually employed in designing underwater large 2-D arrays. The unit vector of the steering direction $\hat{\mathbf{u}}_0$ of a 2-D array can be expressed as [2]

$$\hat{\mathbf{u}}_0 = \left(\sin\alpha_0, \sin\beta_0, \sqrt{\cos\alpha_0^2 - \sin\beta_0^2}\right), \tag{4.1}$$

where α_0 is the angle between the vector $\hat{\mathbf{u}}_0$ and its projection on the plane yz, while β_0 is the angle between the vector $\hat{\mathbf{u}}_0$ and its projection on the plane xz. Analogously, we can define the vector $\hat{\mathbf{u}}$ as

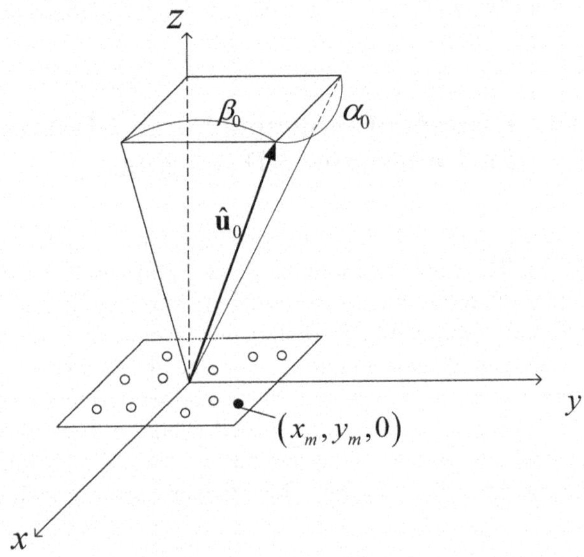

Fig. 4.1 Coordinate geometry and notation for designing underwater large sparse 2-D arrays

$$\hat{\mathbf{u}} = \left(\sin\alpha, \sin\beta, \sqrt{\cos\alpha^2 - \sin\beta^2}\right), \tag{4.2}$$

where the pair (α, β) is the arrival direction. The position vector $\mathbf{v}_m$ of the mth element of the array is denoted by $(x_m, y_m, 0)$. Most of the references on designing narrowband arrays employ narrowband beam pattern under the far-field hypothesis. According to [2], narrowband beam pattern in the far field can be expressed as

$$BP(u, v) = \left|\sum_{m=0}^{M-1} w_m \cdot \exp\left(j\frac{2\pi f}{c}(x_m \cdot u + y_m \cdot v)\right)\right|, \tag{4.3}$$

where M is the number of elements of the array, w_m is the weight coefficient of the mth element, f is the frequency of the transmitted signal, c is the sound speed, and

$$\begin{cases} u = \sin\alpha - \sin\alpha_0 \\ v = \sin\beta - \sin\beta_0 \end{cases}. \tag{4.4}$$

Actually, if we employ an uniform 2-D arrays with the interelement spacing $\lambda/2$, where λ is the wavelength of the frequency f, (4.3) can be written as a discrete Fourier transform (DFT). by descretizing u and v at one frequency bin f. Thus, for unform 2-D arrays, (4.3) can be fast computed by the fast Fourier transform (FFT) [4], which is similar to the time-domain FFT beamforming of (3.25). By using the narrowband beam pattern defined by (4.3), some methods of designing narrowband 2-D arrays were proposed for underwater real-time 3-D acoustical imaging. At present, the methods based on simulating annealing are popularly used, which will be briefly shown in the next section.

4.2.2 Design Based on Simulated Annealing

For narrowband design, the methods rely on the optimization algorithms of simulated annealing [2, 3, 5]. The optimization algorithms need massive iterations of computing array beam patterns. The speed of computing narrowband beam pattern of large 2-D arrays is an important factor influencing the performance of the design methods. These methods [2, 3, 5] find a proper objective function according to the required directivity, and take positions and weights of sensors as optimization variables. Why the optimization algorithms of simulated annealing with massive iterations can be used to design narrowband large 2-D arrays is that narrowband beam patterns of large 2-D arrays can be computed fast.

Trucco et al. [2] proposed the simulated annealing algorithm to design narrowband sparse 2-D array for underwater real-time 3-D imaging, and reduced the element number from 100×100 to 584. Chen et al. [3, 5] modified the method of Trucco et al. [2] with considering both the near-field and far-field directivities, and reduced the element number to 401. The works in [2, 3, 5] realize the sparse design by thinning fully-sampled uniform (on-grid) 2-D arrays. The design methods in [2, 3, 5] are promising to be applied for underwater real-time 3-D acoustical imaging. However, we should note that the hardware cost of the 401 elements is still high.

Diarra et al. [6] employed the simulated annealing algorithm to conduct the design of out-of-grid sparse 2-D arrays, which distinguishes from the sparse 2-D arrays designed on the basis of the uniform (on-grid) 2-D arrays. They demonstrated that the out-of-grid sparse 2-D design was able to obtain fewer elements and better directivity performance. In [6], for the design of thinning the uniform 2-D array, the number of elements is 216; while for the design based on on-grid sparse 2-D array, the number of element is 186. The maximum sidelobe level of the non-grid sparse array is −5 dB lower than that of the standard sparse array designed. The method in [6] is promising for underwater real-time 3-D acoustical imaging.

For designing narrowband large sparse 2-D arrays, other optimization methods such as genetic [7, 8] and particle swarm [9] algorithms were not successful. Currently, due to the computational complexity, the generic and particle swarm algorithms are only employed to optimize arrays of small to moderate size [9–11]. In [12], the number of elements in the fully-sampled uniform 2-D array for the sparse design method based on the genetic optimization is 3209 [12], which is much smaller than that required by our underwater 3-D application, generally 100×100 [1, 2]. Thus, these methods are not applicable for underwater real-time 3-D acoustical imaging at present.

To test the simulated annealing method [2], we conduct an simple design example based on the uniform 2-D array. The number of the uniform 2-D array is 48×48, similar to [47]. The interelement spacing is one wavelength. Because the view of 3-D imaging sonars is usually from −30°w to 30°, the interelement spacing of one wavelength can avoid the grating lobes [30, 47]. The number of elements of our design array is 398. The maximum sidelobe level of our design example is lower than −21 dB. The positions and weights of our design 2-D array is shown in Fig. 4.2. The narrowband beam pattern of the array is shown in Fig. 4.3. Our design example also proves that the methods based on simulated annealing are promising.

The design methods based on simulating annealing [2, 3, 5] are good candidates for underwater real-time 3-D acoustical imaging. Readers who are interested in this topic, can refer to [2, 3, 5, 47]. This book pays more attention to wideband design of large 2-D arrays of underwater 3-D imaging. Thus, the technical details about these methods are not given in this chapter.

4.3 Fast Computation of Wideband Beam Pattern

To date, reports about designing wideband large 2-D arrays by directly using wideband beam pattern for underwater real-time 3-D imaging are rare. The definition of wideband beam pattern is introduced. A fast method of computing wideband beam pattern of large 2-D arrays is given.

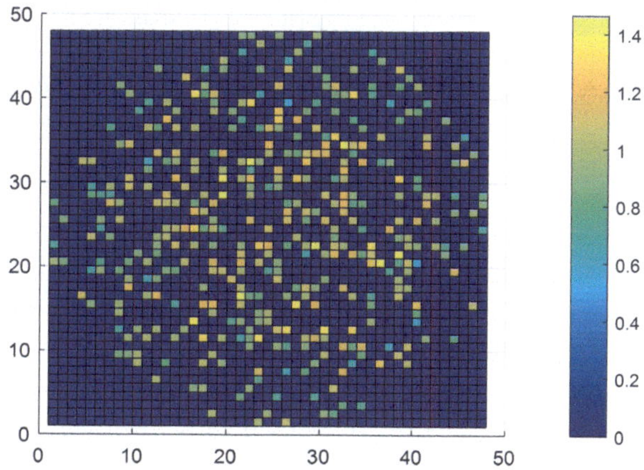

Fig. 4.2 Positions and weights of the elements of our design example based on the simulated annealing [2]

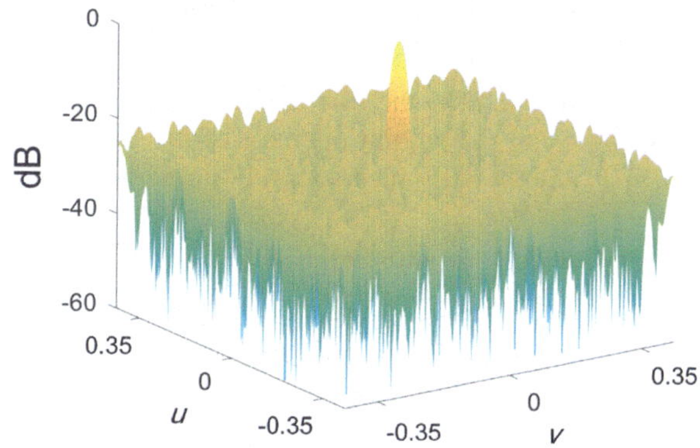

Fig. 4.3 Narrowband beam pattern of our designed sparse 2-D array, as an example

4.3.1 Definition of Wideband Beam Pattern

In underwater real-time 3-D acoustical imaging, the definition of wideband beam pattern of 2-D arrays is different from that of narrowband beam pattern. The coordinate geometry shown in Fig. 4.1 is also used. The wideband beam pattern defines the spatial impulse response firstly [13–16]. Based on (4.1), (4.2) and (4.4), in the far field, the spatial impulse response of a 2-D array is [13]

$$h(u, v, t) = \sum_{m=0}^{M-1} \delta\Big(t - \Big(\frac{x_m u}{c} + \frac{y_m v}{c}\Big)\Big). \tag{4.5}$$

where $\delta(\cdot)$ is the Delta function, u and v are also defined by (4.4). When we denote $s(t)$ as the wideband transmitted or received signal, then the expression of the acoustic pressure [13, 15] can be written as

$$p(u, v, t) = h(u, v, t) * s(t), \tag{4.6}$$

where '*' denotes convolution. Substituting (4.5) into (4.6), (4.6) can be rewritten as

$$p(u, v, t) = \sum_{m=1}^{M} s\left(t - \left(\frac{x_m u}{c} + \frac{y_m v}{c}\right)\right). \tag{4.7}$$

The wideband beam pattern of 2-D arrays at (u, v) is chosen, as that which coincides with the maximum value of the amplitude of the radiated acoustic pressure field over time [13–15], expressed as

$$BP(u, v) = \max_t |p(u, v, t)|. \tag{4.8}$$

4.3.2 Fast Computation Method

In what follows, how the fast method computes wideband beam patterns for 2-D arrays by using the adjoint operator of the 2-D nonuniform fast Fourier transform (NUFFT) and the short inverse fast Fourier transform (IFFT) is shown in the following.

The discrete-time expression of (4.7) is

$$p(u, v, l) = \sum_{m=0}^{M-1} s\left(\frac{l}{f_s} - \left(\frac{x_m v}{c} + \frac{y_m v}{c}\right)\right), \tag{4.9}$$

where f_s is the sampling frequency, and l ($l = 0, \ldots, L-1$) is the discrete-time index. The DFT of (4.9) is given as

$$P(u, v, f_k) = \sum_{m=0}^{M-1} S(f_k) \exp\left(-\frac{i2\pi f_k x_m v}{c} - \frac{i2\pi f_k y_m v}{c}\right), \tag{4.10}$$

where $S(f_k)$ are the DFT coefficients of $s(l/f_s)$, k is the frequency index, and the discrete frequency is given by

$$f_k = k f_s / K \tag{4.11}$$

where K is the length of the DFT. Assume that u and v have P and Q values respectively and are uniformly distributed between -1 and 1 [2, 13, 17]. Then, the respective minimum spacing of u and v is given by

$$\Delta u = \frac{2}{P-1}, \tag{4.12}$$

and

$$\Delta v = \frac{2}{Q-1}. \tag{4.13}$$

The pth value of u and the qth value of v can be given by

$$u_p = \frac{\Delta u}{2} + p\Delta u \quad \left(-\frac{P}{2} \le p \le \frac{P}{2} - 1\right), \tag{4.14}$$

and

$$v_q = \frac{\Delta v}{2} + q\Delta v \quad \left(-\frac{Q}{2} \le q \le \frac{Q}{2} - 1\right) \tag{4.15}$$

respectively. Substituting (4.14) and (4.15) into (4.10), it has

$$P(u_p, v_q, f_k) = \sum_{m=0}^{M-1} S'(f_k) \exp\left(i\left(pa_{m,k} + qb_{m,k}\right)\right), \tag{4.16}$$

where

$$S'(f_k) = S(f_k) \exp(-i\pi f_k x_m \Delta u/c - i\pi f_k y_m \Delta v/c), \tag{4.17}$$

$$a_{m,k} = -2\pi f_k x_m \Delta u/c \tag{4.18}$$

and

$$b_{m,k} = -2\pi f_k y_m \Delta v/c. \tag{4.19}$$

It is found that (4.16) has a form which is similar to the definition of the adjoint 2-D NUDFT (3.46) when $a_{m,k}$ and $b_{m,k}$ of (4.16) are taken as the arbitrary frequencies ω_{1m} and ω_{2m} of (3.46) respectively. Therefore, (4.16) can be computed by using the adjoint operator of the 2-D NUFFT, shown in Chap. 3. Since the distribution of the 2-D array elements is not specifically detailed in the definitions of $a_{m,k}$ in (4.18) and $b_{m,k}$ in (4.19), the computation of (4.16) is suitable for arbitrary 2-D arrays.

After the computation of (4.16) at each frequency points, the inverse discrete Fourier transform (IDFT) is needed to acquire the maximum value of the amplitude of the radiated acoustic pressure field in the time domain to obtain the wideband beam patterns. The IDFT coefficients are written as

$$p(u_p, v_q, l) = \sum_{k=-N_{IDFT}/2}^{N_{IDFT}/2-1} P(u_p, v_q, f_k) \exp\left(\frac{i2\pi lk}{N_{IDFT}}\right), \tag{4.20}$$

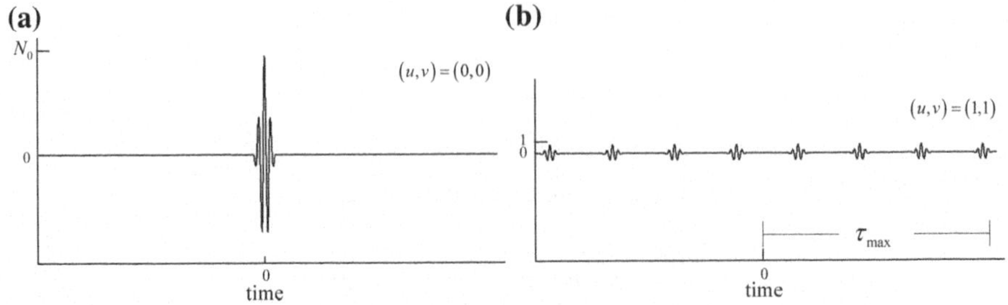

Fig. 4.4 The waveforms of the radiation pattern versus time at **a** $(u, v) = (0, 0)$ and **b** $(u, v) = (1, 1)$

where N_{IDFT} is the length of the IDFT. Then the wideband beam pattern can be given by

$$BP(u_p, v_q) = \max_l \left| p(u_p, v_q, l) \right|. \tag{4.21}$$

Obviously, the length N_{IDFT} influences the computation speed of extracting the maximum values.

Since the number of non-zero discrete frequency points of the excitation signal is crucial to ensure the accuracy and efficiency of the wideband computation in the frequency domain, the length of the DFT and the required points of the IDFT in the fast computation method are discussed in what follows.

Assume that the side length of a square 2-D array aperture is D_1, the central frequency and the nominal bandwidth of the excitation signal $s(t)$ are f_0 and B respectively. The time-domain waveforms of the radiation pattern at $(u, v) = (0, 0)$ and $(u, v) = (1, 1)$ are shown in Fig. 4.4. The waveform at $(u, v) = (1, 1)$ is the longest one among the waveforms of all the values of u and v, the length of which is $2\tau_{\text{max}}$, with τ_{max} being the delay between the two signals radiated by the senor at the apex of the array and the central element. The expression of τ_{max} should be

$$\tau_{\text{max}} = \frac{\sqrt{2}D_1}{2c}. \tag{4.22}$$

To make sure that all maximum values of the amplitude of the radiated acoustic pressure over u and v can be extracted, the length of the DFT N_{DFT} should be larger than or equal to that of the longest waveform at $(u, v) = (1, 1)$. Thus, it has

$$\frac{N_{DFT}}{f_s} \geq 2\tau_{\text{max}}. \tag{4.23}$$

The number of the non-zero frequency points in passband is given by

$$N_f = \frac{B}{f_s/N_{DFT}}. \tag{4.24}$$

In view of (4.22), (4.23) and (4.24), it has

$$N_f \geq \frac{\sqrt{2}BD_1}{c}. \tag{4.25}$$

From (4.25), it can be found that the number of the non-zero frequencies N_f is determined by the bandwidth of the excitation signal B, the 2-D array aperture D_1 and the sound velocity c, while it is irrelevant to the sampling frequency f_s. While for a circular 2-D array with the diameter D_1, similarly, the number of the non-zero frequency points should be

$$N_f \geq \frac{BD_1}{c} \tag{4.26}$$

If D_1 can be described by $N_0\lambda_0$, where λ_0 is the wavelength of the central frequency f_0, (4.25) can be written as

$$\begin{aligned} N_f &\geq \frac{\sqrt{2}BN_0\lambda_0}{c} \\ &= \frac{\sqrt{2}BN_0(c/f_0)}{c} \\ &= \sqrt{2}N_0\left(\frac{B}{f_0}\right). \end{aligned} \tag{4.27}$$

B/f_0 is the relative bandwidth of the excitation signal. Hence, when the array aperture can be described by the number of elements and the wavelength of the central frequency, it can be concluded that the number of the non-zero frequencies N_f is determined by the relative bandwidth and the number of elements.

Usually, to ensure that the waveform obtained by IDFT has the comparable accuracy with that computed by (4.9) in the time domain, N_{IDFT} should be larger than or equal to N_{DFT}. However, since N_{DFT} is usually overlong, it will slow the computation of wideband beam pattern. From (4.21), it can be found that only the time-domain maximum values corresponding to all directions are needed for computing the wideband beam pattern, which means if the maximum values are extracted correctly, the distortion of the time-domain samples caused by the IDFT will be irrelevant to the beam pattern results. Since the waveform envelope is enough to extract the correct maximum values, the passband spectrum centered at f_0 can be shifted to around $f = 0$ to obtain the waveform envelope with the IFFT of N_{IDFT} points, as operated in [18, 19]. N_{IDFT} can be much smaller than N_{DFT} due to the lowering of sampling rate. It is seen from (4.20) that the discrete passband frequency points are around $K_0 = f_0/(f_s/N_{IDFT})$, by which all the discrete passband frequency points

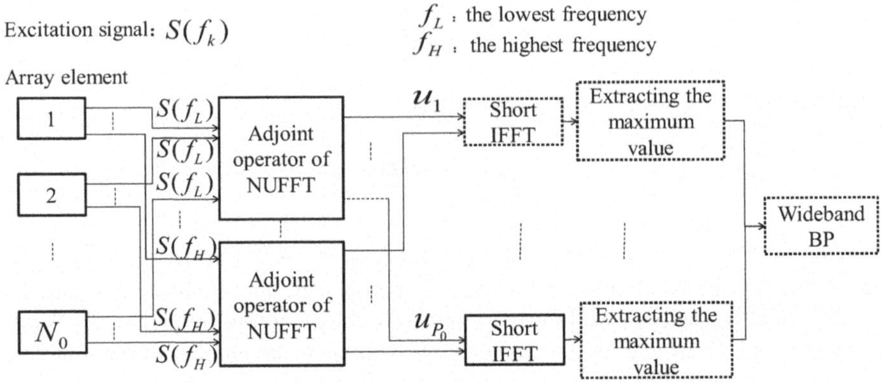

Fig. 4.5 Steps of the proposed fast method of computing wideband beam pattern (BP)

are shifted, the size of the IDFT can be reduced from larger than or equal to N_{DFT} to N_f.

The short IDFT can be expressed as

$$p'(u_p, v_q, l) = \sum_{k=-N_f/2}^{N_f/2-1} P(u_p, v_q, f_{k+k_1}) \exp\left(\frac{i2\pi lk}{N_f}\right), \tag{4.28}$$

where k_1 represents that the k_1 th frequency point is the first non-zero one in the pass band. With the short IDFT, the time-domain envelop of the radiated waveform can be obtained and the maximum value of wideband beam pattern computation can be extracted accurately and efficiently. The wideband beam pattern can be given by

$$BP(u_p, v_q) = \max_l \left| p'(u_p, v_q, l) \right|. \tag{4.29}$$

The short IDFT can be realized by the IFFT with the length N_{IFFT}, where N_{IFFT} is the first power of two, greater than or equal to N_f. In this chapter, the IFFT operation is referred to as the short IFFT, which is applied in order that the time required by extracting the maximum values can be reduced dramatically.

In summary, the proposed fast method of computing wideband beam pattern consists of three steps: (i) transform the acoustic pressure signal in the time domain to the frequency domain according to the number of the non-zero frequency points of the excitation signal; (ii) apply the adjoint operator of the 2-D NUFFT to complete the frequency-domain spatial computation of (4.10); (iii) use the short IFFT to transform the frequency-domain results to the time-domain envelopes and extract the maximum values in the time domain. The proposed fast method is shown in Fig. 4.5.

4.3.3 Performance Evaluation

The proposed method is compared with some existing methods, including the time-domain method, the frequency-domain direct method and the ultrasound simulator Field II [21, 22] through computing wideband beam patterns of 2-D arrays. All the results are obtained on the same personal computer (HP Pro3000 MT E5700) and all the methods are implemented by Matlab. In the implementation, there are no any optimized libraries or accelerating operations such as multithreading, parallel processes being used. Figure 4.6 shows the excitation signal, whose carrier frequency and the bandwidth are 500 and 300 kHz respectively. The expression of the excitation signal is given [20] by

$$g(t) = \begin{cases} g_0 \exp\left[-\frac{\pi^2 \Delta f^2 (t-t_0)^2}{\beta}\right] \sin(2\pi f_0 t) & t \geq 0 \\ 0 & elsewhere \end{cases}, \tag{4.30}$$

where g_0 is a constant, $t_0 = 1.5/\Delta f$, and $\beta = 1.2 \ln 10$. The elements of the arrays serve as point sources. The 2-D array shown in Fig. 4.7 is considered as equally spaced for comparisons. The interspacing between two adjacent elements is $\lambda_0/2$, where λ_0 is the wavelength corresponding to the central frequency f_0.

For the purpose of comparison, the method that computes (4.16) directly in the frequency domain, and then uses the IDFT with the length N_{DFT} to obtain the time-domain maximum values is applied and referred to as the frequency-domain direct method, which is accurate but high in time-consumption. The IDFT is realized by the IFFT, whose length N_{IFFT} is the first power of two which is greater than or equal to N_{DFT}. The steps of the frequency-domain direct method are as follows: complete the computation of (4.16) directly, then apply the IFFT with the length N_{IFFT} to extract the accurate time-domain maximum values.

The wideband beam patterns are also computed in the time domain, which is referred to as the time-domain method. The sampling frequency should be high enough to obtain the accurate results of the time-domain expressions (4.9). The time-domain method adopted computes all the sampled waveforms that every element radiates in all the directions. Unfortunately, computing waveforms on the fly takes too much time. When the wideband 2-D arrays are relatively large, the time cost by the time-domain method may be excessively long. Here, we set a criterion that if the

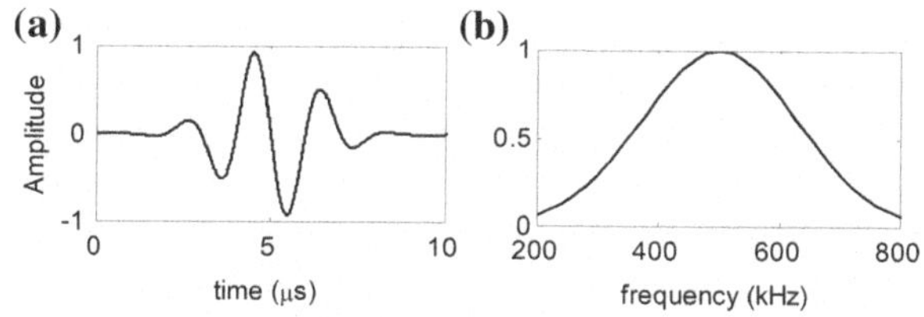

Fig. 4.6 The wideband excitation signal: **a** time-domain waveform; **b** spectrum

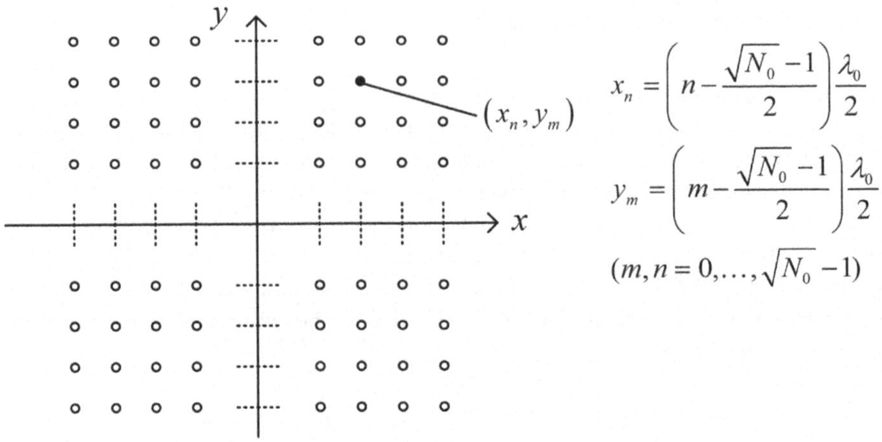

Fig. 4.7 The uniform 2-D array with N_0 elements (the inter-element spacing of the uniform arrays is $\lambda_0/2$)

time exceeds 24 h, namely 8.64×10^4 s, the time-domain method is thought to be failed for computing wideband beam pattern.

To evaluate the accuracy of different methods, the results BP_{DM} of the frequency-domain direct method are taken as reference. The relative error is given by

$$E_r = \left| \frac{BP - BP_{DM}}{BP_{DM}} \right| \times 100\%. \tag{4.31}$$

The sampling frequency of the time-domain method aforementioned is set as 20 MHz, which is larger than the central frequency 500 kHz to guarantee accuracy. For Field II [21, 22], the sampling frequency is 20 MHz and the elements are set to be punctiform to achieve the fast computational speed.

Since the proposed method is an approximated one, a simulation is conducted to evaluate the accuracy of the algorithm. As analyzed in previous sections, the key parameters that determine the computational accuracy and time are the oversampled frequency σ and the interpolated factor J of NUFFT. In order to calculate error with (4.31), the accurate beam patterns needs to be computed with the frequency-domain direct method. A 2-D array with 100×100 elements is chosen, as shown in Fig. 4.7. Table 4.1 present the computational relative errors and time consumptions at different values of σ and J for the 2-D array. From Table 4.1, it can be found that the relative error and the computational time are mainly determined by σ and slightly dependent on J. With the increase of σ, the relative error decreases dramatically and the computational time increases moderately. When $\sigma = 2$, the relative errors become small enough and remain nearly constant (0.15%) with the increase of J, while the computational time increases slightly. Thus, $\sigma = 2$ and $J = 4$ are chosen for the following computation of this book.

Table 4.1 Computational time and relative errors of the proposed method at different values of σ and J for the 2-D array with 100×100 elements

	$\sigma = 1, J = 4$	$\sigma = 1, J = 5$	$\sigma = 1, J = 6$	$\sigma = 2, J = 4$	$\sigma = 2, J = 5$	$\sigma = 2, J = 6$
Time (s)	3.53	3.69	3.70	7.02	7.06	7.11
Error (%)	2.51	1.83	1.66	0.15	0.15	0.15

σ: the oversampled factor; J: the interpolated factor

Table 4.2 Computing time and relative errors of wideband beam pattern using the frequency-domain direct method, the time-domain method, Field II and the proposed method for 2-D arrays

	$N_0 = 8 \times 8$, $P \times Q = 16 \times 16$		$N_0 = 50 \times 50$, $P \times Q = 128 \times 128$		$N_0 = 100 \times 100$, $P \times Q = 256 \times 256$	
	Time (s)	Error (%)	Time (s)	Error (%)	Time (s)	Error (%)
Direct	1.62	0	4.49×10^2	0	8.25×10^3	0
Time domain	12.04	0.58	3.21×10^4	1.12	Failed	Failed
Field II	0.08	0.39	70.35	2.14	11.2×10^2	2.46
Proposed1	0.13	0.04	8.85	0.12	35.30	0.15
Proposed2	0.04	0.04	0.93	0.12	7.02	0.15

Proposed1: the proposed method with only NUFFT; Proposed2: the proposed method using both NUFFT and short IFFT; N_0: the number of elements; $P \times Q$: the number of beams

The next simulation is to evaluate the computational speeds of the proposed method for different arrays. Since the proposed method employs two techniques: NUFFT and the short IFFT, the respective contributions of NUFFT and the short IFFT are separately evaluated. In Table 4.2, Proposed1 represents the proposed method with only NUFFT, and Proposed2 represents the proposed method using both NUFFT and the short IFFT.

Table 4.2 indicates the computing time and the relative errors of wideband beam patterns of 2-D arrays with different number N_0 of elements and different number $P \times Q$ of beams. It is found that for large 2-D arrays, Poposed2 provides about two orders of magnitude faster computing speed compared with Field II, and about three orders of magnitude that compared with the frequency-domain direct method. For the smallest 2-D array with $N_0 = 8 \times 8$, the speeds of the time-domain method, Field II, Proposed1 and Proposed2 are nearly in the same order of magnitude. Compared with that of Proposed1, the computational time of Proposed2 is reduced by about three and five times at $N_0 = 8 \times 8$ and 100×100. It can be seen that the short IFFT helps achieve five-times gain in computational speed for large 2-D arrays. Figure 4.8 shows the wideband beam patterns of using different methods for $N_0 = 100 \times 100$ respectively, which demonstrates that the results of different methods match very well. In addition, when $N_0 = 100 \times 100$, the time-domain method fails because the computational time is excessively long.

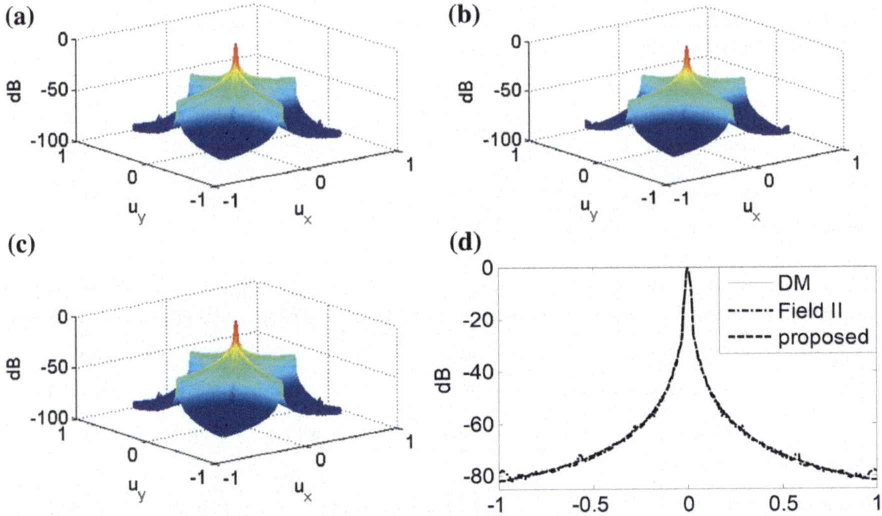

Fig. 4.8 Comparison among the wideband beam patterns obtained by the frequency-domain direct method, the time-domain method, Field II and the proposed fast method for the 2-D array with $N_0 = 100 \times 100$. **a** The direct method. **b** Field II. **c** The proposed fast method. **d** The diagonal line view of the wideband beam patterns (the time-domain method failed because the computational time is excessive)

For the proposed method of computing wideband beam pattern, the time savings are mainly determined by the bandwidth and the array size when the errors' limitation is determined. To reflect the influence of the bandwidth, the simulations with different bandwidths are carried out. For the 2-D array with $N_0 = 100 \times 100$ shown in Fig. 4.7, when the bandwidth is 200 kHz, the computational time of the proposed method is 4.63 s with $\sigma = 2$ and $J = 4$, which is about 2/3 of 7.02 s achieved in the bandwidth of 300 kHz. The comparison verifies that the bandwidth is the key that affects the speed of the proposed method when the errors' limitation is determined.

From Table 4.2, the proposed method shows no obvious advantages for small arrays. For a 2-D array with 8×8 elements and the number of the beams being 16×16, the time consumption of the proposed method is approximate to that of the time-domain method and Field II. The proposed method is superior in saving time when the number of array elements and the number of beams are more than aforementioned values.

At last, a simulation is conducted to verify the feasibility of the fast computation of wideband beam pattern for arbitrary 2-D arrays with the proposed method. A uniform-distribution 2-D random array with 2000 elements [23–25] and a spiral array with 1000 elements [26] are chosen and shown in Fig. 4.9. Owing to the excessively long computing time, the time-domain method cannot be available in this case. Besides, according to the users' guide [22] of Field II, the standard program of Field II cannot compute the wideband beam pattern of the 2-D random and spiral arrays. Thus, the comparisons are carried out between the proposed method and the

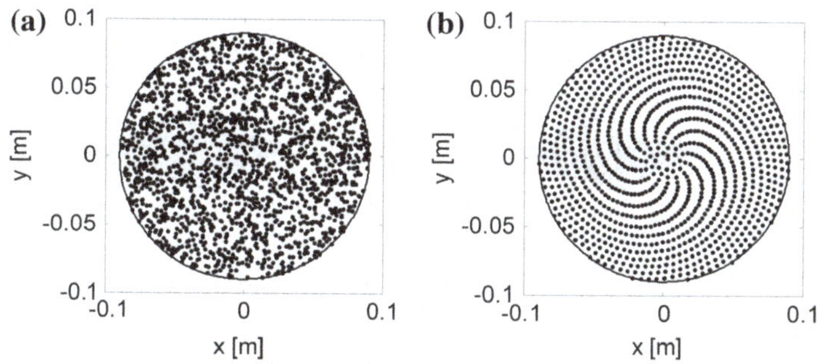

Fig. 4.9 Two non-uniform 2-D arrays: **a** a random 2-D array with uniform distribution and 2000 elements, **b** a spiral 2-D array with 1000 elements

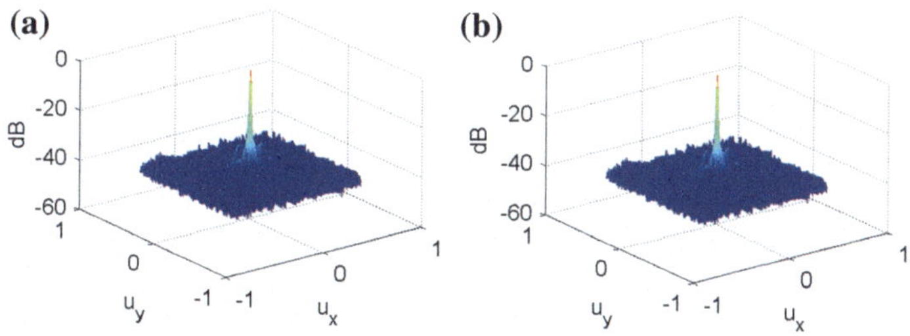

Fig. 4.10 Wideband beam patterns of the random 2-D array for **a** the frequency-domain direct method and **b** the proposed method

frequency-domain direct method. Figure 4.10 shows the wideband beam patterns of the random array computed by the two methods, where the relative error is 0.48%. Figure 4.11 shows the wideband beam patterns of the spiral array computed by the two methods, where the relative error is 0.41%. Both Figs. 4.10 and 4.11 show that the wideband beam patterns obtained by the two methods match perfectly. Moreover, for the random array, the computational time of the proposed method and the frequency-domain direct method is 6.78 s and 1.65×10^3 s respectively. While for the spiral array, the computational time of the proposed method and the frequency-domain direct method is 6.32 s and 1.03×10^3 s respectively. The results for the above two 2-D arrays indicate that the proposed method can be employed to fast compute wideband beam patterns of arbitrary 2-D arrays.

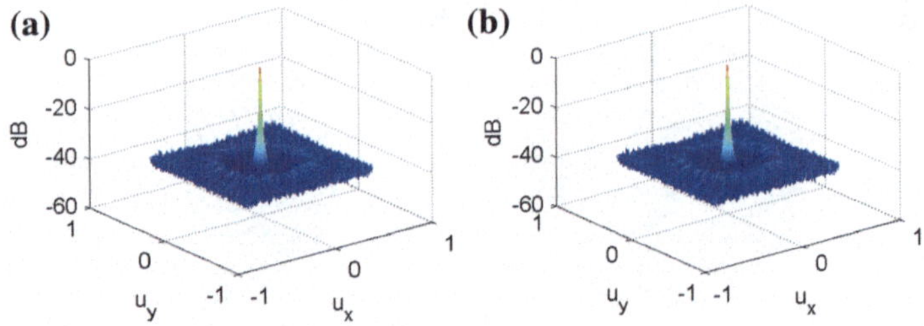

Fig. 4.11 Wideband beam patterns of the spiral array for **a** the frequency-domain direct method and **b** the proposed method

4.4 Wideband Design

4.4.1 Introduction

Designing wideband large 2-D arrays for underwater real-time 3-D acoustical imaging is an open topic. Unfortunately, few papers have discussed this topic. Design of wideband one-dimensional (1-D) arrays has been reported in [17, 20, 27, 28]. In the field of electromagnetism, Spence and Werner utilized aperiodic tilings to design wideband large 2-D arrays [29]. On the basis of Spence and Werner's work, El-makadema et al. [10] proposed two new techniques: controlled randomization and spacing tapering, to obtain better performance. However, the bandwidths of the large 2-D arrays of electromagnetism designed by the above two methods exceed 5 octaves and the sidelobe levels are around −10 dB. According to [1–3, 30], for underwater 3-D real-time acoustical imaging, the typical bandwidth cannot exceed 1 octave and the sidelobe levels required are around −20 dB. The design methods in [29, 38] cannot be used for the underwater application.

4.4.2 Design Method

This chapter shows a method of designing wideband 2-D arrays for underwater real-time 3-D acoustical imaging. Here, it should be pointed out that the relative bandwidth of wideband 2-D arrays in this book is from 10% to 70%. The relative bandwidth which is higher than 70%, is referred to as UWB.

Martínez-Graullera et al. [26] studied the Fermat spiral to design the 2-D arrays and proven to perform well. Owing to the high time consumption of computing wideband beam patterns, the wideband spiral array was designed based on the results of massive iterations of the narrowband beam patterns, which might not guarantee the best performance in the wideband situation. For underwater real-time 3-D imaging, we conduct the wideband design by directly searching the massive wideband beam

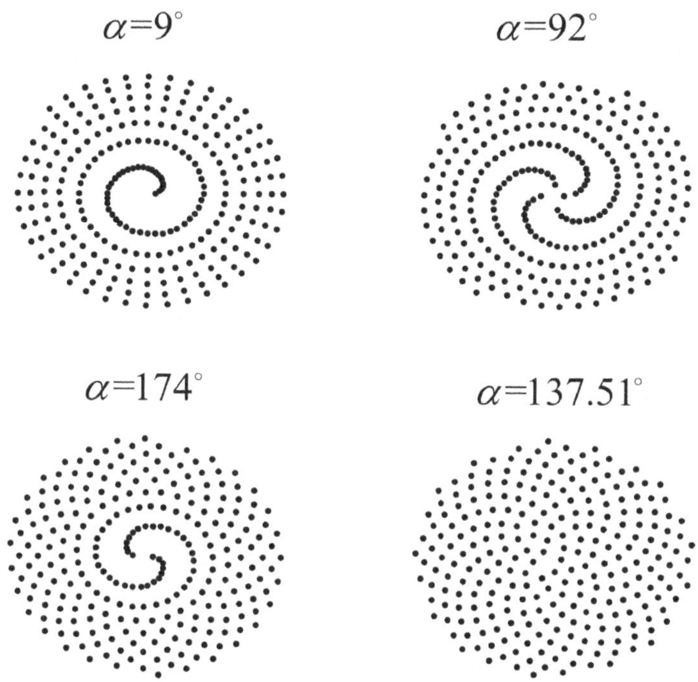

Fig. 4.12 The different Fermat spiral layouts for $\alpha = 9°, 92°, 174°, 137.51°$

patterns on the basis of the fast method of computing wideband beam pattern in Sect. 4.3.

Fermat spiral is defined by the divergence angle α that determines the angular distance between two consecutive elements [26]. The position of the nth element is given by its polar coordinate:

$$\mathbf{R}_n = \left(R_0\sqrt{n\alpha}, n\alpha\right) \quad (n = 0, \ldots, N_e - 1), \tag{4.32}$$

where N_e is the number of elements. R_0 is a constant value to obtain the desired aperture diameter D_2, which is defined as

$$R_0 = \frac{D_2}{2\sqrt{(N_e - 1)\alpha}}. \tag{4.33}$$

Figure 4.12 presents different spiral layouts with different divergence angles ($\alpha = 9°, 92°, 116°, 137.51°$) and $N_e = 500$. Due to the diversity of the spiral layouts, optimization by directly searching in all the layouts can help achieve a better-performance 2-D array.

In our design, we set that the relative bandwidth $Q = 50\%$, $N_e = 500$ and $D_2 = 60\lambda_0$. The criterion of determining the best divergence is minimizing the maximum sidelobe level of the wideband beam pattern. The same configuration as that used by Martínez-Graullera et al. [26] is applied to do direct search on all the values of the divergence angles $\alpha = 0° : 0.01 : 180°$ based on wideband beam pattern, where 18,000 iterations of computing wideband beam patterns are needed. The total time of direct search using the proposed fast method of computing wideband beam pattern is 1.76×10^5 s.

4.4.3 Performance of the Designed Array

To evaluate the designed wideband array, of which the wideband beam pattern is compared with that of other three arrays which have same bandwidth and apertures: a fully sampled array (FULL) with circular aperture which has the diameter $60\lambda_0$ and 11,305 elements, and a uniform random sparse array (Random) with 500 elements. For comparisons, the uniform random array [23–25] with 500 elements has been designed by taking the best of 18,000 cases in view of the wideband beam pattern. The random array has a circular aperture with the diameter $60\lambda_0$ shown in Fig. 4.13b. The positions of the 500 elements of the random array are uniformly randomly distributed in the circular aperture. The steps of designing the random array are given as follows: (i) generate 18,000 cases of uniform random array (The reason of choosing 18,000 is that the number of spirals needed by the proposed method of design wideband 2-D arrays is also 18,000.); (ii) compute the wideband beam patterns of the 18,000 cases; (iii) find the best case which has the minimum sidelobe level of wideband beam pattern. The similar method of obtaining random arrays has also been used in [26].

Figure 4.13a–f is a summary of the comparison of the four arrays' results. Figure 4.14 presents the lateral profiles along u-axis of the wideband beam patterns with $v = 0$. From Fig. 4.14, it can be seen that the main lobes above −20 dB for the three arrays almost overlap with each other. The maximum sidelobe levels for the random array and designed arrays are −26.20 and −32.49 dB respectively. Table 4.3 shows that all the 2-D arrays offer nearly the same results in lateral angle resolution, because the lateral angle resolution is determined by the size of the active aperture. As for the sidelobe level which determine the imaging quality, the designed array achieves the best performance.

Table 4.3 Performance of different 2-D arrays

2-D arrays	Lateral resolution − 3 dB (°)	Maximum sidelobe level (dB)
FULL	0.95	Null
Random	0.95	−26.20
Designed	0.95	−32.49

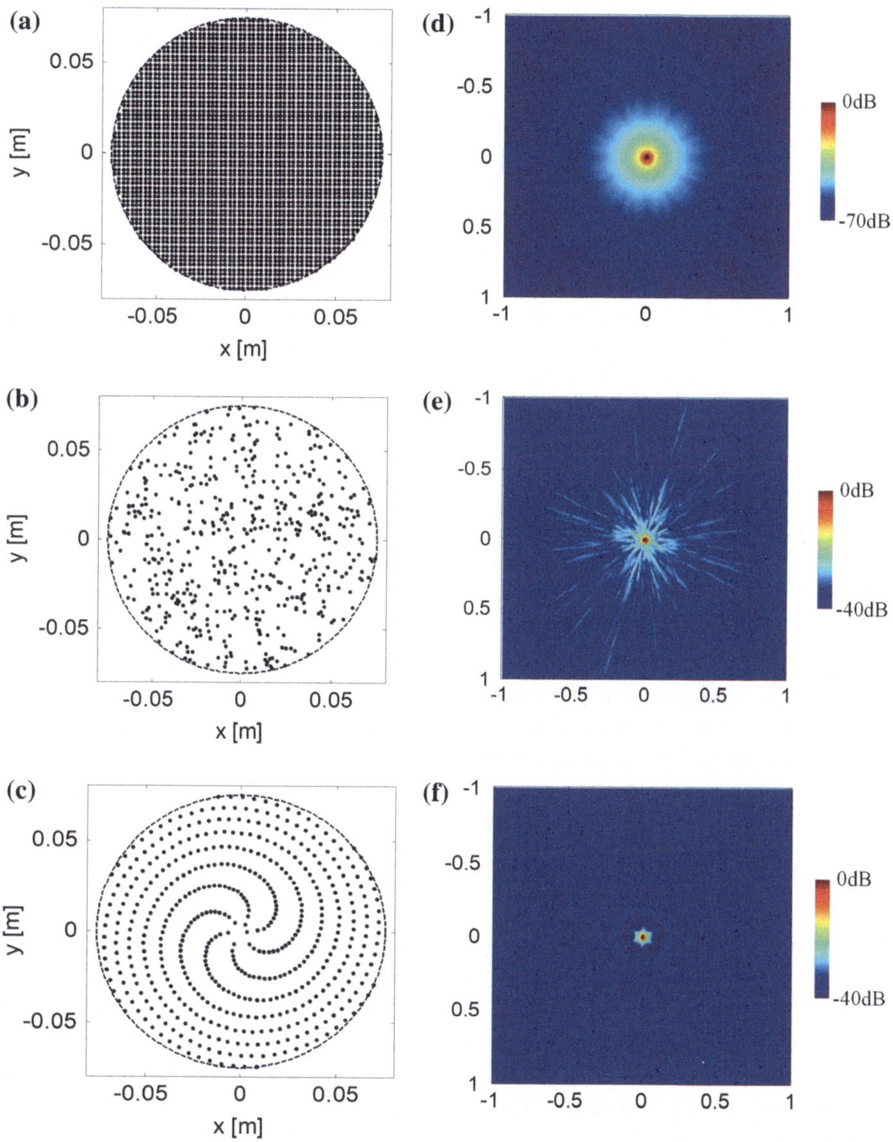

Fig. 4.13 Layouts and wideband beam patterns for different arrays: **a**, **d** for the fully sampled uniform array with circular aperture; **b**, **e** for the uniform random array; **c**, **f** for the designed Fermat spiral

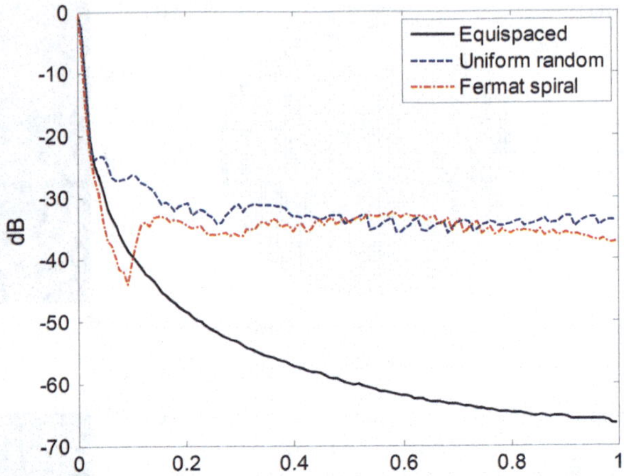

Fig. 4.14 Lateral profiles of the related 2-D wideband beam patterns over 80-degrees

It is aforementioned that the arrays designed by using narrowband beam patterns cannot guarantee the best performance in wideband. To verify the fact, we calculates the wideband beam pattern of the array designed by using narrowband beam patterns. The steps of narrowband design are the same with that of the proposed wideband design method, except the use of narrowband beam patterns. The best divergence angle of using narrowband beam patterns is 91.29°. The maximum sidelobe level of using narrowband beam patterns is −30.87 dB, which is higher than −32.49 dB obtained by using the proposed method based on wideband beam patterns.

4.5 UWB Ultrasparse 2-D Arrays

The UWB technique can help to achieve a ultralow hardware cost. How to apply UWB arrays to underwater real-time 3-D acoustical imaging is introduced. The feasibility of using the UWB technique to underwater 3-D imaging is analyzed first. Then, the directivity of UWB ultrasparse 2-D arrays is evaluated. How to apply the modulation excitation technique to improve the SNR is also explained. Finally, simulations are given to test the performance of UWB ultrasparse 2-D arrays in underwater 3-D imaging.

Before introducing, the definition of UWB for the underwater needs to be clarified. Conventional narrowband signals are with a small fractional bandwidth $\Delta f/f_0 \ll 1$, where Δf is the bandwidth and f_0 is the carrier frequency. For typical radar applications, references [31, 32] define UWB as signals which have relative bandwidths equal to or larger than 100%. For medical ultrasound imaging, UWB is defined as the signal which has a pulse duration shorter than three cycles of the central frequency [14]. For underwater real-time 3-D acoustical imaging, references [33, 34] have pointed out that a relative bandwidth equal to or larger than 70% allows the directivity of UWB 2-D arrays to meet the imaging requirements. The definition of UWB signals in this book is given as

$$\frac{\Delta f}{f_0} \geq 0.7. \tag{4.34}$$

Additionally, to facilitate the description, an impulse signal of (4.30) is employed.

4.5.1 Feasibility of Using the UWB Technique for Underwater 3-D Imaging

Absorption attenuation is unavoidable when acoustical signals propagate in underwater channels. A logarithmic absorption coefficient (to the base 10) $\alpha(f)$ has been given to assess the absorption in [35–37], which depends on frequency f. As the frequency grows higher, the absorption coefficient becomes larger, which means that the UWB acoustical signals will distort when they propagate a certain distance underwater. Therefore, to apply the UWB technology into underwater real-time 3-D acoustical imaging, the absorption attenuation must be assessed.

Assume that the acoustical signals are plane waves, the transmitted intensity at the frequency f is $I_0(f)$, and the distance between the imaged target and the transmitter is r. The received intensity at the location of the transmitter [35] is

$$I_1(f)=I_0(f) \cdot 10^{-2r\alpha(f)/10}. \tag{4.35}$$

When $g(t)$ of (4.30) is taken as the transmitted acoustical signal, according to (4.35), the spectrum of the received signal $g_1(t)$ should be

$$\begin{aligned} G_1(f) &= G(f) \cdot \left(10^{-2r\alpha(f)/10}\right)^{1/2} \\ &= G(f) \cdot 10^{-r\alpha(f)/10}, \end{aligned} \tag{4.36}$$

where $G(f)$ is the spectrum of $g(t)$.

A simple and specific expression of $\alpha(f)$ given in [35] is chosen, which is

$$\alpha(f)=\frac{0.1 f^2}{1+f^2}+\frac{40 f^2}{4.100+f^2}+2.75 \times 10^{-4} f^2+0.003, \tag{4.37}$$

where $\alpha(f)$ is in dB per kiloyard, and f is the frequency in kilohertz. The range unit used in this paper is meter. One yard is equal to 0.914 m. To keep the consistency of the range units, the absorption in dB per meter is given as

$$\alpha'(f)=\frac{\alpha(f)}{914}. \tag{4.38}$$

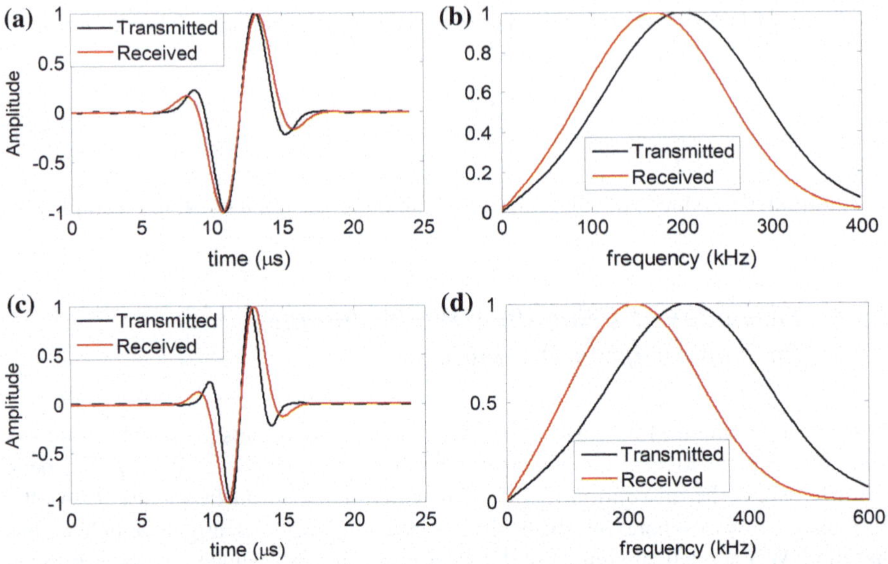

Fig. 4.15 Waveforms and spectra of the transmitted and received signals for different parameters: **a**, **b** for $\Delta f / f_0 = 1$, $f_0 = 200$ kHz, and $r = 200$ m; **c**, **d** for $\Delta f / f_0 = 1$, $f_0 = 300$ kHz, and $r = 200$ m (the time delays of the received signals have been compensated)

By using (4.36), (4.37) and (4.38), it can be assessed whether the received signal is UWB or not.

When the attenuation caused by absorption in homogenous seawater is considered, for $\Delta f / f_0 = 1$, $f_0 = 200, 300$ kHz, and $r = 200$ m, the transmitted and received acoustical waveforms and their spectra are shown in Fig. 4.15. It can be found that although the mean frequencies of the received signals decrease, the received signals are still UWB.

To reflect the general law of UWB signals for underwater real-time 3-D acoustical imaging, enough numeric calculations are conducted with different values of f_0 and r. Assume that the mean frequency and bandwidth of the received signal $g_1(t)$ are denoted by f_0^r and Δf^r. The fractional bandwidth $\Delta f^r / f_0^r$ is employed to judge whether the received signal is UWB or not.

In addition, if the mean frequency f_0^r decreases too much, the angular and range resolutions of the underwater 3-D imaging system will be influenced severely. Thus, the decrease of f_0^r must be considered to judge whether the UWB signals can be applied to underwater real-time 3-D acoustical imaging in a certain situation or not. In this book, the ratio f_0^r / f_0 is used to assess the decrease of f_0^r. Even though the received signals are UWB, we propose that when $f_0^r / f_0 < 0.7$, they are not considered as applying to underwater real-time 3-D acoustical imaging systems.

In the following, the relative bandwidths of the transmitted UWB signals are set as 1. Calculations that check $\Delta f^r / f_0^r$ changes with the carrier frequency f_0 for different distances are conducted and the results are shown in Fig. 4.16, which proves that

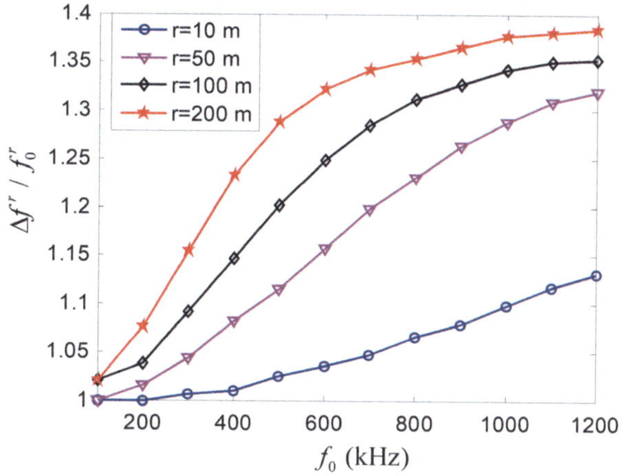

Fig. 4.16 Relationship between the fractional bandwidth of the received signal and the carrier frequency of the transmitted signal for different values of r

the fractional bandwidths of all the received signals are greater than or equal to 1. Hence, according to (4.34), all the received signals are UWB.

Relationship between the mean frequency f_0^r of the received signal and the carrier frequency f_0 of the transmitted UWB signal for different distances are studied, as shown in Fig. 4.17. It can be found that the mean frequency f_0^r of the received signal varies distinctly with the variation of f_0 for each r. For example, in the case of $r =$ 10 m and $f_0 = 1200$ kHz, f_0^r/f_0 is 0.75, which means the UWB signal is valid for underwater 3-D imaging; while for $r = 200$ m and $f_0 \geq 400$ kHz, f_0^r/f_0 is lower than 0.70, which means the UWB signal is not valid for underwater real-time 3-D imaging. The general criteria of usable UWB signals in underwater real-time 3-D acoustical imaging in Fig. 4.17 are summarized as follows: for $r = 10$, 50, 100, and 200 m, the carrier frequency f_0 of the UWB transmitted signal can arrive at 1200, 600, 400, and 300 kHz, respectively. Based on the above analysis, it can be concluded that applying the UWB technology into underwater real-time 3-D acoustical imaging is feasible.

4.5.2 *Directivity of UWB 2-D Arrays*

Consider that an ideal UWB array with N elements at arbitrary locations is used to receive a very short impulse whose amplitude is 1. Because the impulse is very short, all the spacings between every two adjacent elements exceed the impulse length. The received impulses of the elements are incoherent in other directions except the steering direction of the ideal UWB array. Therefore, the amplitude of the coherent output impulse of the ideal UWB array at the steering direction should be N; while the maximum amplitude of the output waveform of the ideal UWB array in other

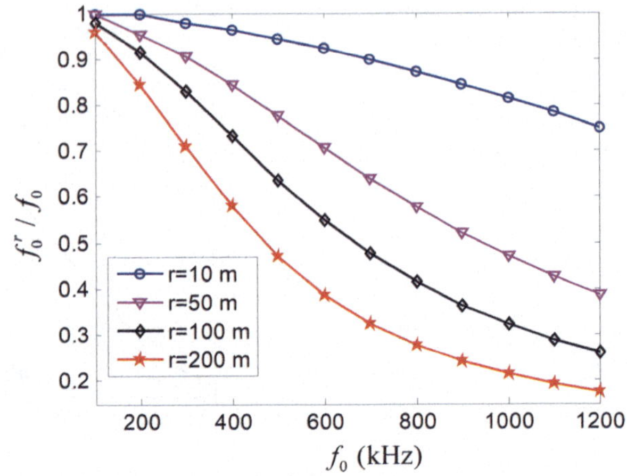

Fig. 4.17 Relationship between the mean frequency of the received signal and the carrier frequency of the transmitted signal for different imaging ranges r

directions should be 1. As a result, for the ideal UWB array, the ideal maximum sidelobe level [14] is given by

$$SL_{ideal} = 20\lg\frac{1}{N}. \tag{4.39}$$

The maximum sidelobe level required in underwater real-time 3-D imaging systems is around −22 dB [1–3, 30, 33, 34]. According to (4.39), for $SL_{ideal} = -22$ dB, the ideal number of elements should be

$$\begin{aligned} N_{ideal} &= \frac{1}{10^{SL_{ideal}/20}} \\ &= \frac{1}{10^{-22/20}} \\ &\approx 13. \end{aligned} \tag{4.40}$$

It means that in the ideal UWB situation, 13 elements can achieve the −22 dB sidelobe level required in the underwater application. However, in practical situations, since the target scatterers may not be so distinct that an ideal UWB array is hard to obtain, the maximum sidelobe level is usually higher than the ideal one. Thus, in the design of the practical UWB arrays, the number of elements should be larger than the ideal number of elements computed by (4.39).

Currently, this book mainly focuses on discussing the capacity of UWB arrays in underwater real-time 3-D acoustical imaging rather than designing an optimal UWB array. We just provide two ultra-sparse arrays to verify the advantages of UWB arrays here [33].

One array shown in Fig. 4. 18a is an equally spaced square array with the inter-element spacing $10\lambda_0$, of which the number of elements is 7×7 and the side length

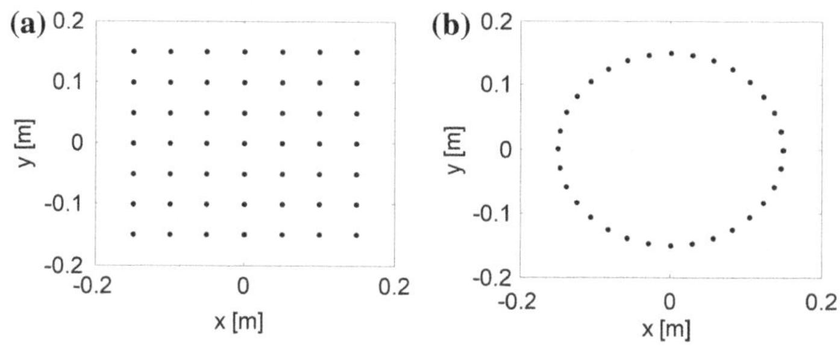

Fig. 4.18 Two ultra-sparse arrays: **a** the equally spaced square array; **b** the annular array

is $60\lambda_0$. The other array shown in Fig. 4. 18b is an annular uniform array with 32 elements and the diameter $60\lambda_0$. The carrier frequency f_0 of the array is set 300 kHz.

Figure 4.19a–f show the beam patterns of the two arrays at the fractional bandwidth of $\Delta f/f_0 = 10, 50, 100\%$. Because the steering and the observation angles have been represented by normalized angular variables u and v, naturally the steering angles are corresponding to $u = 0$ and $v = 0$, i.e. $\beta = \beta_0$ and $\varphi = \varphi_0$ in Fig. 4.19. It can be seen that the sidelobe levels of the beam patterns of both arrays decrease greatly with the increase of the fractional bandwidth. For the square array, the high grating lobes still appear at $\Delta f/f_0 = 100\%$, namely in the UWB situation, as shown in Fig. 4.19c, which is mainly due to the element shadowing [14, 34]. While for the annular array, when $\Delta f/f_0 = 100\%$, no evident grating lobes appear and the sidelobe levels are below -23 dB, as shown in Fig. 4.19f, which meet the sidelobe level requirement -22 dB for the underwater 3-D imaging application. Therefore, in view of sidelobe level requirement, the UWB annular array with 32 elements is feasible for underwater real-time 3-D acoustical imaging.

For underwater 3-D acoustical imaging, an equally spaced half-wavelength square array with the side length $60\lambda_0$ will have 121×121 elements; while for the UWB array, 32 elements or fewer may achieve the similar sidelobe level performance. Hence, in the underwater 3-D imaging application, UWB arrays have the potential to achieve an ultra-sparse structure, which means the ultra-low hardware cost.

In the above analysis, we only use the transmitted signal to compute the beam pattern, where the absorption attenuation is not considered. In fact, when the signal distortion caused by the absorption attenuation occurs, the beam pattern of the array also changes. To assess the absorption attenuation's influences on the beam pattern of the UWB annular array, the beam pattern is also computed by using the received signal shown in Fig. 4.15c. The result is depicted in Fig. 4.20, where the steering angles are corresponding to $u = 0$ and $v = 0$ as usual. It is seen that the sidelobe levels are still below -23 dB. Hence, when the absorption attenuation is taken into account, the UWB annular array can still be applied to underwater real-time 3-D imaging.

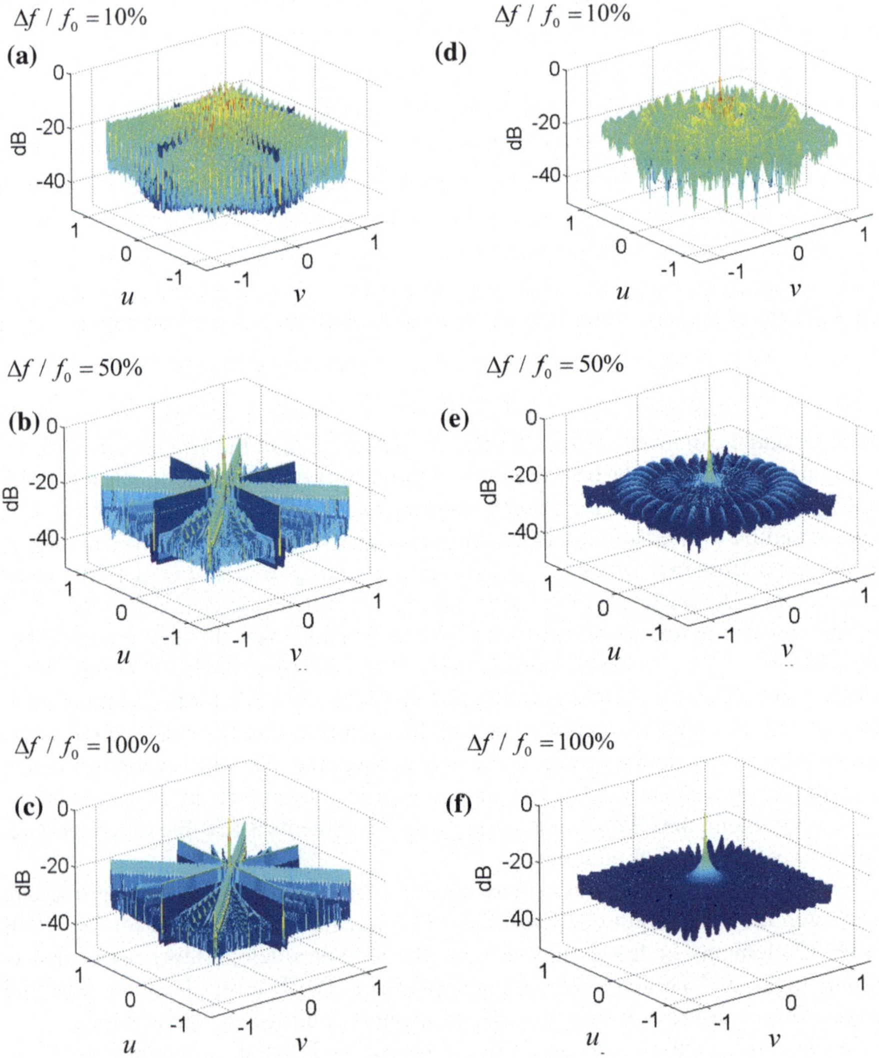

Fig. 4.19 Beam patterns of the two arrays at the fractional bandwidth of $\Delta f/f_0 = 10, 50, 100\%$: **a–c** for the equally spaced square array with 7×7 elements; **d–f** for the annular array with 32 elements

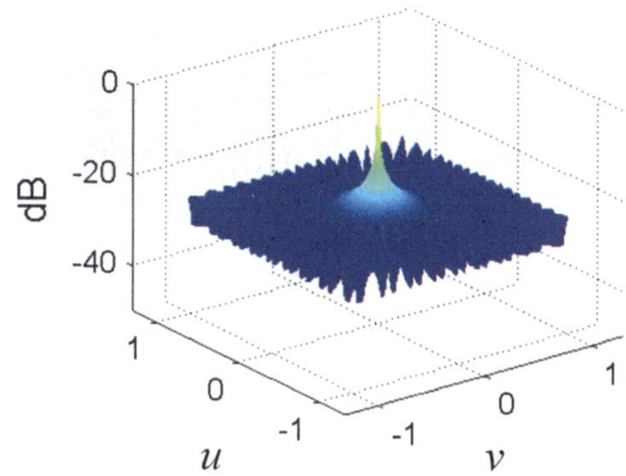

Fig. 4.20 Beam pattern of the annular array in Fig. 4.18b with considering the absorption attenuation of $r = 200$ m

4.5.3 Modulation Technique for Improving SNR

The modulated excitation technique consists of modulated excitation signal design and pulse compression [38]. A modulated excitation UWB signal and mismatched filter are chosen for underwater 3-D real-time acoustical imaging. The practicability of the modulated excitation technique in the underwater channel situation is demonstrated, while the negative effect of using modulated excitation technique, i.e. the computational load of the pulse compression, is also discussed.

The gain of modulated excitation signals can be measured by the time-bandwidth product (*TB*) [38]. There are many filters that can be used to pulse compression, such as matched [39], mismatched [40], inverse [41], and Wiener filters [42], among which the matched filter achieves the optimal SNR gain. When using the matched filter, the ideal SNR gain [38] is

$$GSNR = 10\lg TB. \tag{4.41}$$

Many kinds of modulated excitation signals such as linear frequency modulation (FM) signals [41], nonlinear FM signals [43], binary complementary Golay codes [44] have been applied to radars, medical ultrasound imaging. In [40], it has been shown that linear FM signals have the best performance and most robust features for medical ultrasound imaging. Since the underwater 3-D acoustical imaging situation is similar to the medical ultrasound imaging one, the linear FM signals are chosen as the modulated excitation in this book, which can be expressed in complex notation [40] as:

$$\psi(t) = \exp\left\{ j2\pi\left[\left(f_0 - \frac{B}{2}\right)t + \frac{B}{2T}t^2\right]\right\}, \quad 0 \le t \le T. \tag{4.42}$$

A linear FM signal with $f_0 = 300$ kHz, $B = 400$ kHz and $T = 100$ μs is shown in Fig. 4.21a.

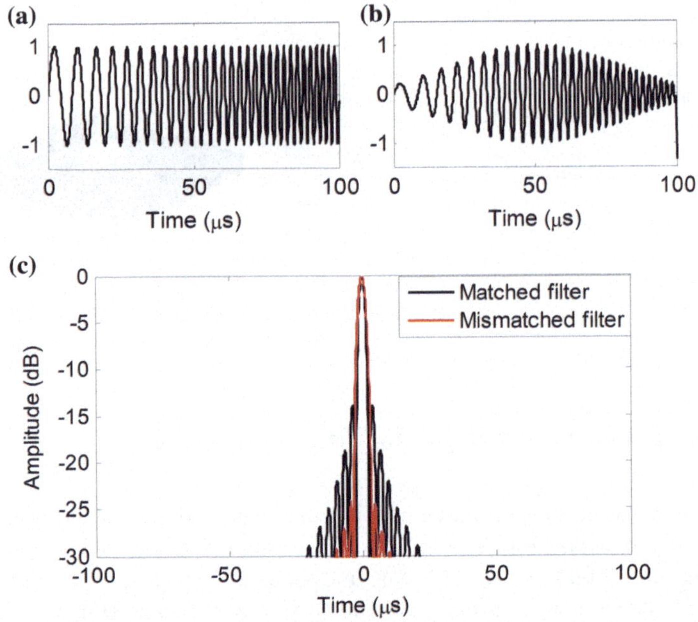

Fig. 4.21 Linear FM signal designed for underwater 3-D imaging and the pulse compression filter: **a** the linear FM signal; **b** the impulse response of the pulse compression filter, namely the mismatched filter; **c** the outputs of the matched (black line) and mismatched (red line) filters

For underwater real-time 3-D acoustical imaging, the sidelobe level of beam patterns is usually set −22 dB [30, 33]. Thus, the output range sidelobe level of the pulse compression filters should not be higher than −22 dB to achieve a high-quality 3-D image. Since the matched filter can only provide a range sidelobe level of −13.6 dB, as given in Fig. 4.21c, a mismatched filter should be employed for underwater real-time 3-D acoustical imaging.

A usual approach to realize mismatched filters is to use a window function on the time-domain response of the matched filter [40]. In this chapter, a Dolph-Chebyshev window with −30 dB sidelobe level is used to compress the range sidelobe levels of pulse compression. The time-domain response of the designed mismatched filter is shown in Fig. 4.21b. Because the mismatched filter helps compress the range sidelobe levels with the decrease of the system bandwidth, the bandwidth of the original linear FM signal is set 400 kHz to make the system keep UWB. From Fig. 4.21c, it can be found that the range sidelobe level of the mismatched filter reaches −24 dB, which is lower than −22 dB.

In the above analysis, the effect of the underwater absorption attenuation has not been discussed. To evaluate the performance of the mismatched filter in the frequency-dependent absorption medium, with estimation of (4.37), the received FM signal is also simulated at the imaging range $r = 200$ m. The simulation results is that the range sidelobe level of the mismatched filter becomes lower and reaches below −30 dB when the underwater absorption attenuation is taken into account, which is better than that of not considering absorption. At the same time the gain of SNR, i.e. *GSNR*, decreases from 14.7 to 10 dB.

The fast frequency-domain digital beamforming methods such as the chirp zeta transform (CZT) [18] and the NUFFT [45], have been proven to be proper choices for underwater real-time 3-D acoustical imaging. If the computational load of the pulse compression is higher than that of the fast beamforming methods, it will be questioned whether the modulated excitation technique can be applied to underwater real-time 3-D acoustical imaging or not. Thus, the computational load of the pulse compression should be evaluated and compared with that of the fast beamforming methods.

As shown in [18, 45], when using those frequency-domain beamforming methods to realize 3-D acoustical imaging, three steps are necessary. The first is to convert the segmented time-domain samples into frequency domain by FFT. The second is to use the CZT or NUFFT to realize the fast beamforming. The final step is to convert the frequency-domain beam signals to the time domain by IFFT. Therefore, when the modulated excitation technique is applied to underwater 3-D imaging, the pulse compression of UWB underwater 3-D imaging systems can be realized in the frequency domain following the first step and before the second step.

Assume that $M(f_k)$ is the discrete frequency-domain response of the mismatched filter, where f_k represents the kth frequency bin, and the DFT coefficients of the received signal of the nth element is $S_n(f_k)$. The frequency-domain pulse compression of using the mismatched filter [38] can be described by

$$CS_n(f_k) = S_n(f_k) \cdot (M(f_k))^*, \tag{4.43}$$

where $(\cdot)^*$ represents the conjugate operation. Equation (4.43) shows that at one frequency bin, the pulse compression needs 1 complex multiplication, namely 6 real operations without distinguishing the real addition and multiplication. The total number of real operations of the pulse compression at one frequency bin for N elements can be expressed as

$$B_{PC} = 6 \cdot N. \tag{4.44}$$

The CZT beamforming [18] can be applied to equally spaced 2-D arrays or sparse arrays thinning from the equally spaced 2-D arrays. When a 2-D array with $N_e \times N_e$ elements is employed and $M_b \times M_b$ beams need to form, the number of real operations of the CZT beamforming at one frequency bin [18] is

$$B_{CZT} = 6\left[N_e^2 + M_b^2 + L^2\right] + 20L^2 \log L, \tag{4.45}$$

where L is a power of two, greater than or equal to $N_e + M_b - 1$. Moreover, the number of real operations does not reduce with the decrease of elements when using the sparse technique.

Table 4.4 Computational load comparison

Pulse compression	CZT beamforming	NUFFT beamforming
0.96×10^5	1.77×10^{10}	1.18×10^{10}

The NUFFT beamforming [45] is suitable for arbitrary 2-D arrays. When a 2-D with N elements is employed, $M_b \times M_b$ beams need to form, then the number of real operations of the NUFFT beamforming at one frequency bin [45] is

$$B_{NUFFT} = N\left(8J^2 - 2\right) + 5M_b'^2 \log_2\left(M_b'^2\right), \tag{4.46}$$

where J is the interpolation coefficient, $M_b' = \sigma M_b$, and σ is the oversampled factor. For the underwater application, $J = 4$ and $\sigma = 2$ are the proper choices [45].

In the following, a numerical comparison is given. The parameters for the comparison are summarized as:

- $N = 32$;
- $N_e \times N_e = 121 \times 121$;
- $M_b \times M_b = 256 \times 256$;

Assume there are $N_f = 500$ frequency bins in the passband. The computational loads of the pulse compression, the CZT and NUFFT beamforming in the whole passband are shown in Table 4.4. It can be found that the computational load of the pulse compression is five orders of magnitude lower than that of the CZT or NUFFT beamforming. It means that comparing with that of the beamforming methods in underwater real-time 3-D acoustical imaging, the computational load of the pulse compression can be neglected. Therefore, the pulse compression is feasible and practical for underwater real-time 3-D acoustical imaging to improve the SNR of the system.

4.5.4 *Simulation for Ultrasparse UWB 3-D Imaging*

The simulation of the backscattered acoustical signals is conducted by using the method given in [46], which will be shown in Chap. 5. An underwater 3-D scene with three scattering square frames (A, B and C) shown in Fig. 4.22 is employed, of which the positions are given by the coordinate system in Fig. 4.1. The square frame A represents the near field and the high SNR case. The square frame B represents the far field and the high SNR case. While the square frame C is in the far field and the low SNR case.

Figures 4.23 and 4.24 depict the 2-D slices of the 3-D images at different focusing distances but in high SNR cases obtained by using the designed UWB 3-D imaging system. It can be found that both the square frames can be seen clearly. Figure 4.25a, b depict the 2-D slices in the case of the far field and the low SNR without and

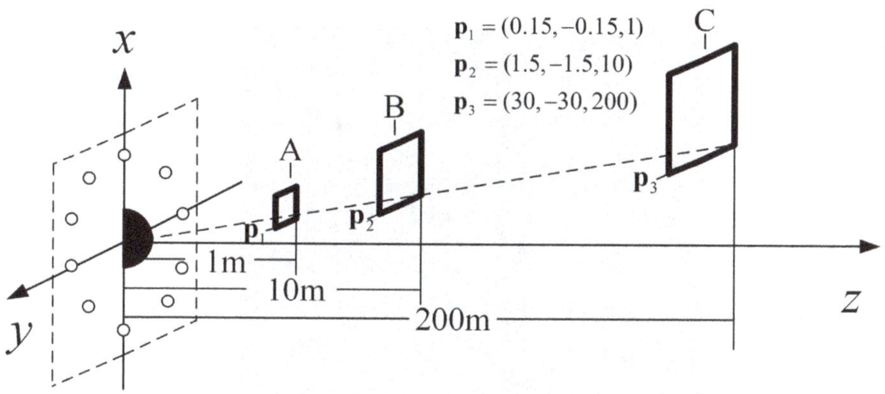

Fig. 4.22 A 3-D scene of the UWB imaging sonar with the three scattering square frames (A, B and C) (the three scattering square frames parallel to the plane *xy*. The side length of A, B and C are 0.075 m, 0.75 m and 15 m. The positions of them are described by the point coordinates $\mathbf{p}_1$, $\mathbf{p}_2$ and $\mathbf{p}_3$)

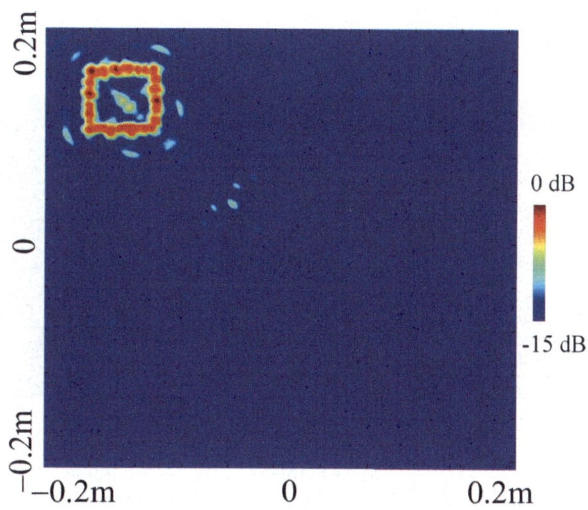

Fig. 4.23 2-D slice of the 3-D image of the square frame A ($z = 1$ m) when using the designed UWB imaging system prototype

with using the modulated excitation techniques respectively. In Fig. 4.25a where the modulated excitation technique is not applied, the square frame nearly disappears. However, in Fig. 4.25b where the modulated excitation technique is applied, although the image quality worsens in the low SNR situation, the square frame can still be seen clearly. The comparison between Fig. 4.25a, b verifies that for UWB ultrasparse underwater 3-D acoustical imaging, the modulated excitation technique can help increase the SNR and improve the imaging quality.

To sum up, the results in Figs. 4.23, 4.24 and 4.25b demonstrate that the designed UWB 3-D acoustical imaging system prototype has a good performance.

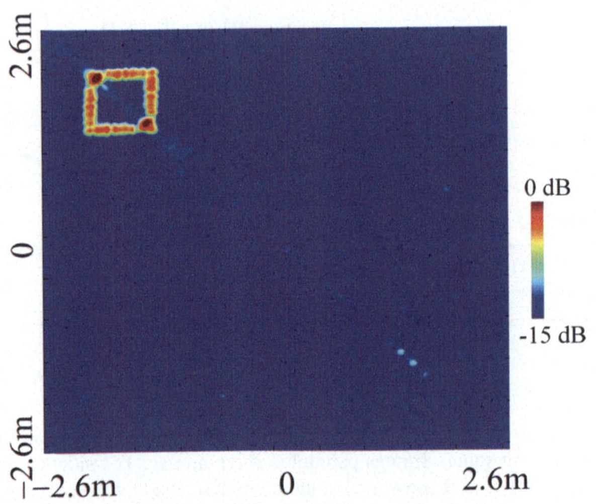

Fig. 4.24 2-D slice of the 3-D image of the square frame B ($z = 10$ m) when using the designed UWB imaging system prototype

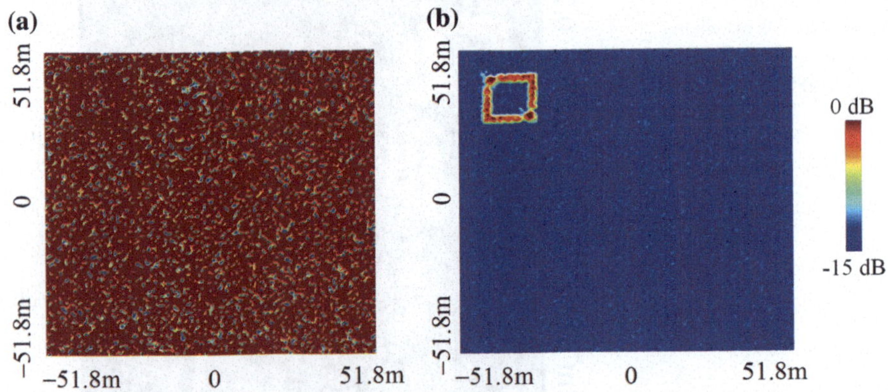

Fig. 4.25 2-D slice of the 3-D image of the square frame C ($z = 200$ m) when using the designed UWB imaging system prototype: **a** without using the modulated excitation technique; **b** using the modulated excitation technique

4.6 Ultralarge Ultrasparse UWB 2-D Arrays

It is possible to use ultralarge UWB ultrasparse 2-D arrays to obtain a high angular resolution of 0.1°. The concept, directivity and imaging simulation of ultralarge UWB ultrasparse 2-D arrays are shown in this chapter.

4.6.1 Concept

This chapter proposes the concept of ultralarge UWB ultrasparse 2-D arrays for high-resolution underwater 3-D acoustical imaging through analyzing the array resolution and beam steering. 'Ultra-large' is for achieving high resolution. 'Ultra-sparse' is for the feasibility of implementation. 'Ultra-sparse' will cause grating lobes at the narrowband scenario, while UWB signal is used to average this effect. Therefore, 'ultra-wideband' is essential to realize 'ultra-sparse'.

It is also known that for a 2-D circular array of the central frequency f_0 with the diameter D_3 and the corresponding wavelength λ_0, the angular resolution θ_{res}, i.e. the main-lobe width of the beam pattern [14], is determined by

$$\theta_{res} \approx \frac{\lambda_0}{D_3}. \tag{4.47}$$

It means that the larger the diameter D_3, the higher the angular resolution. Therefore, the angular resolution is substantially decided by the aperture size of arrays, and nearly irrelative to the bandwidth. Based on (4.47), the lateral resolution will be

$$r_{lateral} \approx l\theta_{res}, \tag{4.48}$$

where l is the imaging distance.

As shown in [1, 2, 30, 47], for the most existing underwater 3-D imaging systems, the angular resolutions are around 1°. According to (4.48), for the imaging distance 100 m, the lateral resolution is about 1.7 m, which is a value much larger than the range resolution of a wideband 2-D array and is also too large to image small targets in this distance. Suppose that the angular resolution is enhanced to 0.1°, i.e. the lateral resolution is about 17 cm in 100 m distance, the imaging quality will be significantly improved. Therefore, we focus on discussing the angular resolution being around 0.1°, where D_3 should be enlarged to $600\lambda_0$ according to (4.47). Few papers have discussed the design of 2-D arrays with the aperture size around $600\lambda_0$, as well as 2-D arrays of such aperture size having been considered in any a practical underwater 3-D acoustical imaging system. Compared with those having the 'large' aperture size of around $50\lambda_0$, the 2-D arrays with aperture size over $500\lambda_0$ are 'ultra-large', we refer to them as 'ultra-large' arrays in this book.

As an example, for a fully-sampled uniform squared array with the side size $600\lambda_0$ and the half-wavelength inter-element spacing, the number of elements will be 1200×1200, of which the hardware cost is unimaginable for implementation in any a practical imaging system. The sparse design of the ultra-large arrays is hence mandatory. In general, a maximum of hundreds of sampling channels connected with hundreds of elements is the most cost-effective scheme which is realizable with current techniques [30, 47]. If the element number of the ultra-large 2-D array can be decreased from 1200×1200 to several hundreds, it will be affordable for the system implementation. Since the sparsity of the element number decreasing from $1200 \times$

1200 to around 100 is about 4 orders of magnitude, it is referred to as 'ultra-sparse' in this book. Evidently, 'ultra-sparse' design is necessary for using the ultra-large 2-D array to achieve a $0.1°$ angular resolution.

Currently, the existing sparse design techniques [2, 3, 5] can thin 2-D arrays having about ten thousands elements and decrease the number of elements of large 2-D arrays from several ten thousands to several hundreds. However, few papers have discussed what kind of 2-D arrays should be adopted and how to decrease the element number of ultra-large 2-D arrays from over one million to about or less than one hundred. In this chapter, the ultra-sparse design method of ultralarge UWB 2-D arrays is proposed and demonstrated through the design of an annular ultralarge ultrasparse UWB 2-D array with 100 elements, which reduces the element number by 4 orders of magnitude.

As analyzed previously, using an UWB array is the essential mean to realize the sparsity, since it has many advantages compared with a narrowband one. For the UWB array with the number of elements N_e, for ultra-sparse it should have

$$\Delta\bar{d} \geq cT, \tag{4.49}$$

where $\Delta\bar{d}$ is the average inter-element spacing, and T is the pulse duration. If an uniformly-spaced ultra-sparse array is adopted, the ideal sidelobe level of the beam pattern can be given by (4.39).

In practice, whether the ideal sidelobe level can be obtained by an UWB array depends on the imaging medium [14]. If there are only some distinct scatterers which means the imaging medium is ultra-sparse, the ideal sidelobe level can be achieved. However, the targets in an underwater acoustical imaging scene may be diverse and complicated. In this case, the medium integrates the energy in the beam waveform in the time domain over the entire duration, the sidelobe level of an UWB array will become [14]

$$SL_{sub-ideal} = 10\log\frac{1}{N_e}. \tag{4.50}$$

In summary, the concept of ultralarge ultrasparse UWB 2-D arrays illustrates that an 'ultra-large' 2-D array can help realize the high-resolution 3-D imaging, an 'ultra-sparse' design is necessary to reduce the element number of the array from over one million to around 100, and an 'ultra-wideband' scheme is to provide the essential base to enable the 'ultra-sparse' design. We propose the concept of ultralarge ultrasparse UWB 2-D arrays and gives the design method through designing a specific annual ultralarge ultrasparse UWB 2-D array. Through the fast beamforming algorithm of arbitrary 'ultra-large' 2-D-array [18, 45], 3-D images of targets are obtained with the designed annual ultralarge ultrasparse UWB 2-D array. The problem of finding the optimal ultra-sparse structure will not be discussed in this book.

4.6.2 Directivity of Ultralarge Ultrasparse UWB 2-D Arrays

The huge number of elements of the 'ultra-large' 2-D arrays decides that the optimization of the structures of 2-D arrays is impossible with the traditional design methods such as simulated annealing et al., due to the tremendous computational load of computing beam pattern of 'ultra-large' 2-D-arrays. As a compromise, the projection slice method [14, 48] is applied in this book to find the sub-optimal structures of the 2-D arrays. As shown in [14, 49], the projection slice method can be adopted to analyze 2-D arrays through a set of equivalent linear arrays. By projecting all element excitations onto a rotation of the x-axis at the angle φ on the plane of the 2-D array, the equivalent one-dimensional (1-D) array aperture is obtained [14]. In both the near and far fields, the radiation pattern of the 2-D array at the specific φ cut is the same as that of the equivalent 1-D array. Figure 4.26 gives an example of three equivalent 1-D arrays of a 9-element square-grid array by using the projection slice method. In Fig. 4.26a, b, there exist element overlaps which are referred to as element shadowing. It means that the equivalent 1-D arrays have less than 9 effective elements (3 for Fig. 4.26a and 5 for Fig. 4.26b). Both of the cases result in the sidelobe level losses. Obviously, the equivalent 1-D arrays which have the most elements will achieve the lowest sidelobe level, as shown in Fig. 4.26c. To achieve nearly ideal performance at all φ cuts, the equivalent 1-D array should have the least shadowing at as many φ cuts as possible. In theory, the optimal performance can be obtained via the periodic arrangement of 1-D array in all φ cuts. For simplification, some sub-optimal schemes can be obtained by using the least shadowing criterion in practice. As illustrated in [14], curvilinear structures such as ring and spiral 2-D arrays may keep somewhat periodic equivalent projected 1-D arrays. Therefore, the projection slice method provides an efficient way to choose the sub-optimal structures of the 2-D arrays.

As analyzed above, an ideal sparse 2-D array should have minimum element shadowing when projecting to φ cuts. The statement "somewhat periodic" means

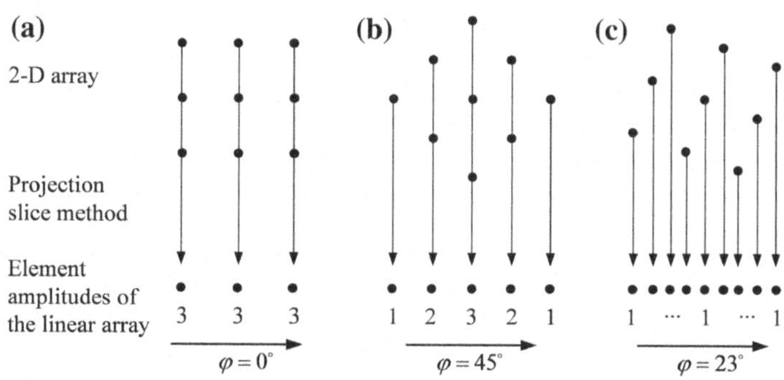

Fig. 4.26 Three equivalent 1-D arrays obtained from the 9-element 2-D array by using the projection slice method. Cuts at $\varphi = 0°, 45°, 23°$

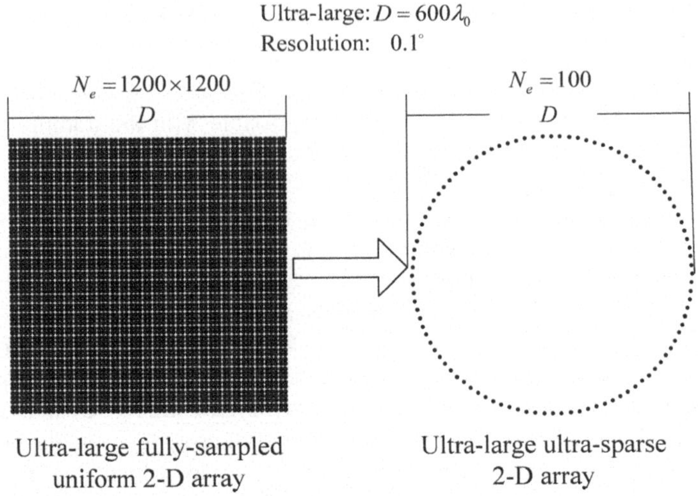

Fig. 4.27 Schematic diagram of realizing the angular resolution 0.1° using a 100-element ultralarge ultrasparse UWB 2-D annular array to replace the ultra-large fully-sampled uniform 2-D array

the projected aperiodic 1-D arrays should be somewhat periodic to achieve a good performance. As illustrated in [14], ring and spiral 2-D arrays may keep somewhat periodic equivalent projected 1-D arrays. Thus, it is thought that annular 2-D arrays may have relatively weaker element shadowing and all the projected equivalent 1-D arrays may maintain somewhat periodic. The advantages of annular 2-D arrays have been mentioned in [14]. In this chapter, for employing an ultra-large 2-D array to obtain the angular resolution 0.1°, we propose to use the annular 2-D structure as a sub-optimal scheme.

Having chosen the annular structure of the ultralarge ultrasparse UWB 2-D arrays, the number of the required elements of the annular ultralarge ultrasparse UWB 2-D array should be determined. As shown in [1–3, 30], the required sidelobe level of 2-D arrays for high quality underwater 3-D imaging should be around −20 dB. According to (4.50), if −20 dB $SL_{sub-ideal}$ is needed, the number of elements should be 100. It means when applying the UWB technique into underwater 3-D imaging, 100 elements can satisfy the sidelobe levels requirements in theory.

Figure 4.27 presents the ultra-sparse scheme of realizing the angular resolution 0.1° using a 2-D annular array with 100 elements to replace the ultra-large fully-sampled uniform 2-D array.

Analyzing with the projection slice method, the worst element-shadowing case of an annular 2-D array in Fig. 4.27 will occur at the axial symmetric cases, as sketched in Fig. 4.28. In any other case, the element shadowing will be weaker. Moreover, even in the worst case, with half elements being shadowed, the sidelobe level loss will be only 3 dB according to (4.50), which is still superior to uniform 2-D arrays. Because for the uniform 2-D array with 1200 × 1200 elements, 1200 × 1199 elements will be shadowed in the 4 worst cases, of which the sidelobe level loss reaches 61 dB.

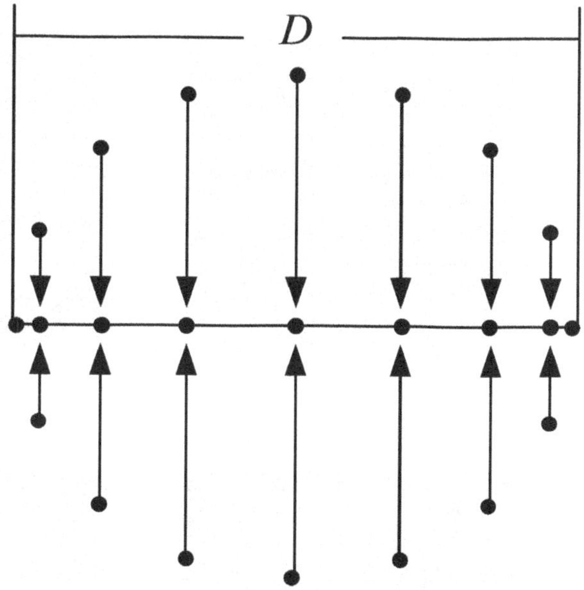

Fig. 4.28 The worst case of element shadowing of an annular 2-D array, nearly half elements are shadowed to form an equivalent 1-D array

To facilitate the discussion of bandwidth, the excitation signal is expressed by (4.30). With the fast computation of wideband beam pattern of an arbitrary large 2-D array in this chapter, the beam patterns of the annular ultralarge ultrasparse UWB 2-D array with the aperture $600\lambda_0$ and 100 elements are obtained and evaluated in different relative bandwidths. Figure 4.29a–d show the beam patterns with the relative bandwidth $B = 10$, 40, 70 and 100% respectively.

From Fig. 4.29, it can be seen that the main-lobe width (central region) in all cases are the same (0.1° as designed), while the sidelobe levels decrease with the increase of the relative bandwidth. Since the main lobes in Fig. 4.29a–d are too narrow, the performance of the beam patterns in the region around the respective main lobes will be further evaluated and compared by using the 2-D slices in the below.

To evaluate the performance of the annular ultralarge ultrasparse UWB 2-D array in the worst case, the beam patterns of the projected equivalent 1-D array which keeps the worst element-shadowing (half elements are shadowed) are computed in different relative bandwidths. Since the beam pattern of the projected equivalent 1-D array is equivalent to that of the 2-D slices of Fig. 4.29, similar as the description of the 2-D arrays, the x axis of the beam patterns of the equivalent 1-D arrays in Figs. 4.30, 4.31, 4.32, 4.33 is also described with u.

Figure 4.30 illustrates the beam patterns of the projected equivalent 1-D array in the relative bandwidth $B = 10, 40, 70$ and 100%. From Fig. 4.30a, it can be found that for $B = 70\%$ and $B = 100\%$, the maximums of sidelobe levels are below −30 dB. Exactly, a −32 dB maximum sidelobe level is obtained when B = 70%, which is low enough for underwater 3-D acoustical imaging. Although there exists a little

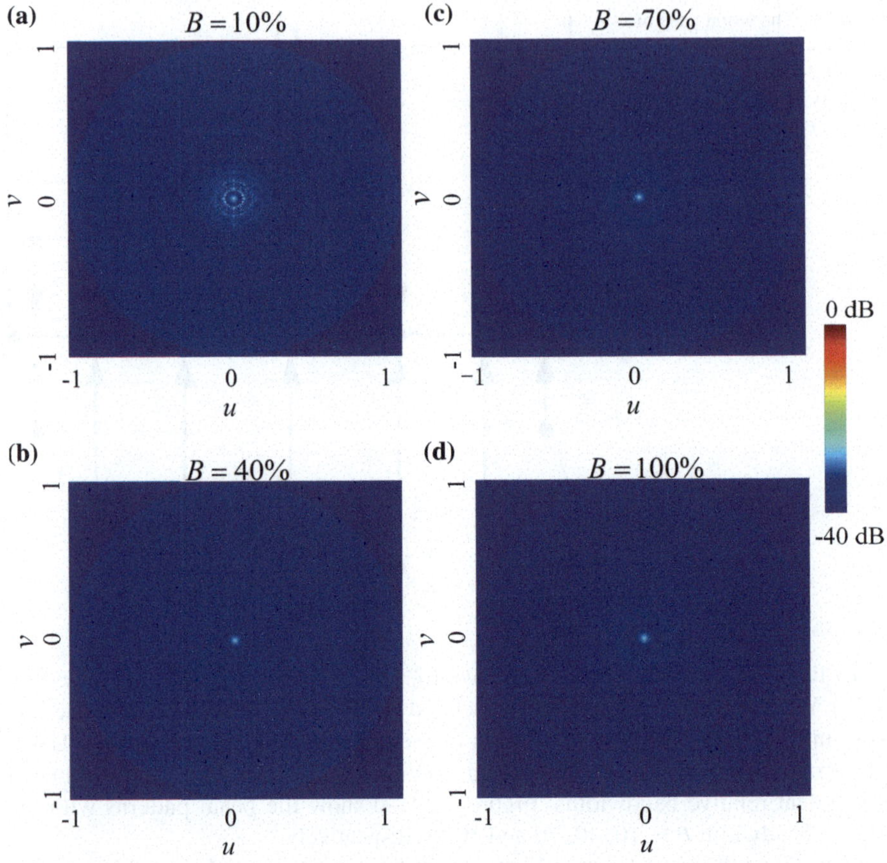

Fig. 4.29 Beam patterns of the annular 2-D array with the aperture diameter $600\lambda_0$ and 100 elements in relative bandwidths of: **a** $B = 10\%$; **b** $B = 40\%$; **c** $B = 70\%$; **d** $B = 100\%$

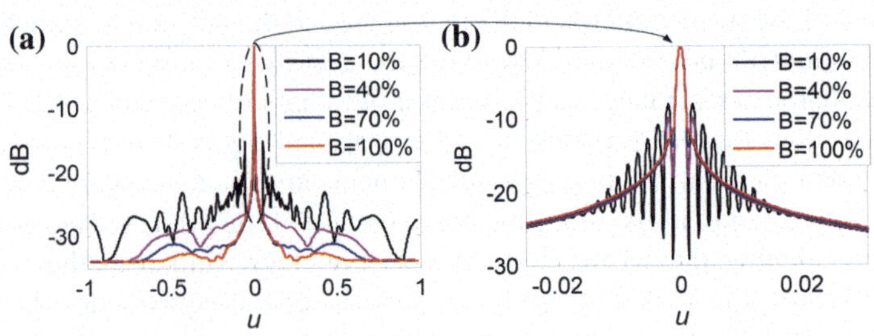

Fig. 4.30 Beam patterns of the projected equivalent 1-D array of the annular 2-D array in the worst element-shadowing case in the relative bandwidths of $B = 10, 40, 70$ and 100%: **a** the beam patterns for $-1 < u < 1$; **b** the beam patterns around the main lobes

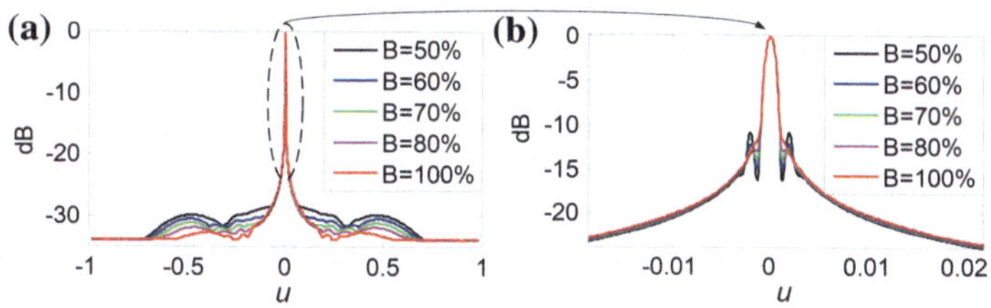

Fig. 4.31 Beam patterns of the projected equivalent 1-D array of the annular 2-D array in the worst element-shadowing case in the relative bandwidths of B = 50, 60, 70, 80 and 100%: **a** the beam patterns for $-1 < u < 1$; **b** the beam patterns around the main lobes

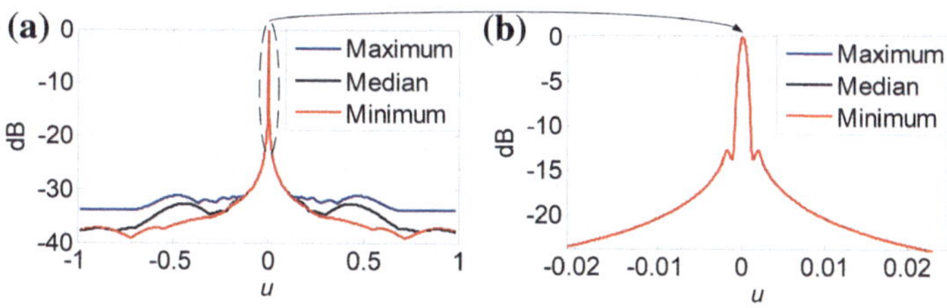

Fig. 4.32 Beam patterns of 3 projected equivalent 1-D arrays which achieve the maximum, median and minimum sidelobe levels of the annular ultralarge ultrasparse UWB 2-D array in $B = 70\%$, respectively. **a** The beam patterns for $-1 < u < 1$; **b** the beam patterns around the main lobes

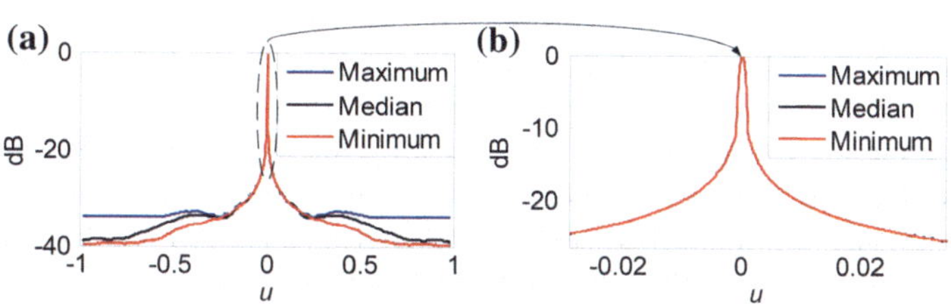

Fig. 4.33 Beam patterns of 3 projected equivalent 1-D arrays which achieve the maximum, the median and the minimum sidelobe levels of the annular ultralarge ultrasparse UWB 2-D array in $B = 100\%$, respectively. **a** The beam patterns for $-1 < u < 1$; **b** the beam patterns around the main lobes

fluctuation in the main lobe when $B = 70\%$, as shown in Fig. 4.30b, it is lower than 1 dB and usually can be neglected in imaging application [14].

Figure 4.31 compares the beam patterns of the projected equivalent 1-D array in larger relative bandwidths, i.e. B = 50, 60, 70, 80 and 100%. It further verifies that when $B \geq 70\%$, all the beam patterns of the annular 2-D arrays have similar performance and all the maximums of sidelobe levels are below −30 dB, the bigger the relative bandwidth, the lower the maximum sidelobe level.

Figure 4.32 presents the beam patterns of 3 projected equivalent 1-D arrays which achieve the maximum, the median and the minimum sidelobe levels of the annular ultralarge ultrasparse UWB 2-D array in $B = 70\%$, respectively. From Fig. 4.32a, it can be found that all the sidelobe levels for the maximum, the median and the minimum sidelobe level cases are relatively flat and all the maximums of sidelobe levels are below −30 dB, among which the difference is little. Moreover, from Fig. 4.32b, it can be found that the main lobes for the 3 cases are completely overlap. The results mean that the designed annular 2-D array in $B = 70\%$ has the near optimal performance in sidelobe levels. Figure 4.33 presents the similar beam patterns as Fig. 4.32 but in $B = 100\%$. Compared with Fig. 4.32, it can be found that there are no evident difference between the beam patterns for the $B = 70\%$ and $B = 100\%$ cases, except a little bit lower sidelobe levels for $B = 100\%$. Therefore, the relative bandwidth 70% is a proper value to decide the UWB for the high resolution 3-D imaging.

Based on the above analyses, it can be concluded that when $B \geq 70\%$, the annular ultralarge ultrasparse UWB 2-D array with 100 elements can satisfy the beam-steering requirement of underwater 3-D acoustical imaging and can achieve the angular resolution around 0.1°. It can be thought as an UWB array in underwater imaging if the relative bandwidth is larger than or equal to 70%.

4.6.3 *Simulation for High-Resolution UWB Underwater 3-D Imaging*

The above analyses have evaluated the beam patterns of the annular ultralarge ultrasparse UWB 2-D array in different relative bandwidth. An imaging simulation of complicated targets is conducted to further testify the performance of the designed annular ultralarge ultrasparse UWB 2-D array in underwater 3-D imaging.

An underwater 3-D acoustical imaging scene is simulated and the imaging is conducted through the beamforming with the annular ultralarge ultrasparse UWB 2-D array. The relative bandwidth of the transmitted signal is set 70%. The frequency

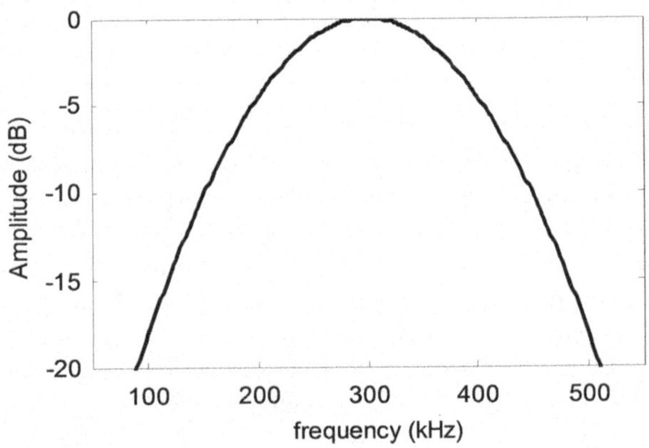

Fig. 4.34 Frequency response of the receiving transducer

response of the receiving transducer is given in Fig. 4.34, of which the −6 dB relative bandwidth is also 70%. The central frequency is 300 kHz.

To testify the performance of the proposed annular ultralarge ultrasparse UWB 2-D array, an uniformly-spaced ultralarge ultrasparse UWB 2-D array is employed, shown in Fig. 4.35. The interelement spacing is $60\lambda_0$. The element number is 11×11.

The underwater 3-D scene is shown in Fig. 4.36. Four scattering sticks (A, B, C and D) are employed for imaging, of which the length and width are all 1.5 and 0.1 m respectively. Four scattering sticks are placed in two groups on a plane paralleling to the plane *xy*. A and B are mutually orthogonal, while C and D are parallel to each other. The spacing between C and D is 0.25 m, corresponding to about a 0.3° angular distance. The radius of all the sticks is 0.05 m. The positions of two groups are described by their corner-point coordinates $\mathbf{p}_0$ and $\mathbf{p}_1$ respectively. Here, it should be noted that the imaging targets are located in the near field of the ultralarge ultrasparse UWB 2-D array. Currently, according to [1–3, 30], the far-field beam pattern is generally employed to evaluate the directivity of the 2-D arrays. As shown in [18], underwater 3-D acoustical imaging systems should work both in both the near and far field. In fact, the near-field imaging is more complicated, where the dynamic focusing technique needs to be used [18, 34]. In fact, if the near-filed imaging result is good, the far-field one should be better. Therefore, in this section, the near-field scenario is considered.

The backscattered acoustical signals of the complicated targets are simulated by using the method in [46]. All sticks are set to be made of steel and the surfaces of them are rough. The surfaces of all the sticks consist of massive small scatterers, with a density of 1×10^6 scatterers/m^2. The radius of the scatters is around 0.08 mm. Similar with [46], to simulate the roughness, the height of the scatterers on the stick surfaces is perturbed by a Gaussian variable with a standard deviation of 1.2 mm and a zero mean.

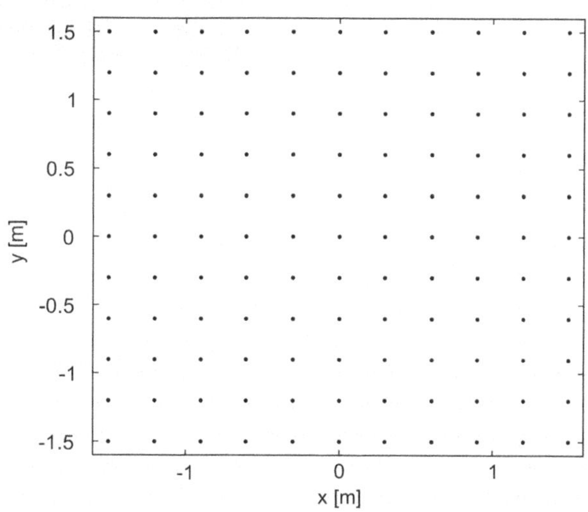

Fig. 4.35 The uniformly-spaced ultralarge ultrasparse UWB 2-D array with the interelement spacing $60\lambda_0$ and 11×11 elements

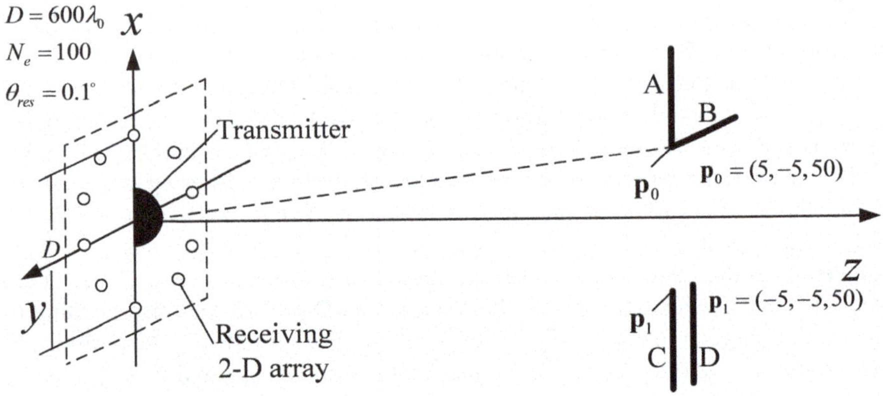

Fig. 4.36 An underwater 3-D acoustical imaging scene of using the annular ultralarge ultrasparse UWB 2-D array with 100 elements

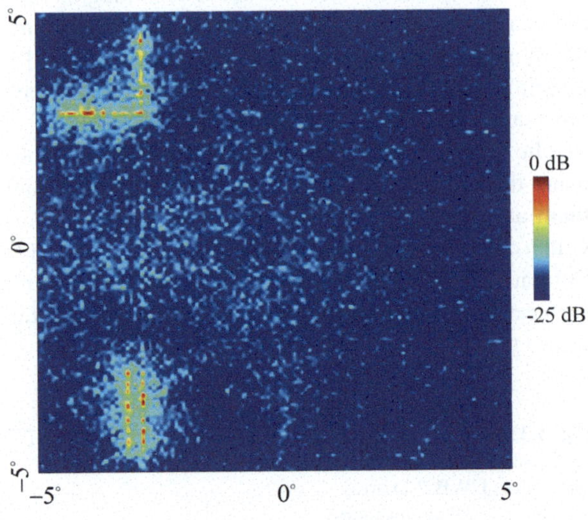

Fig. 4.37 Front view of the 3-D image obtained by using the annular ultralarge ultrasparse UWB 2-D array with the aperture diameter $600\lambda_0$ and 100 elements

Since we focus on improving the imaging angular resolution and lowering the hardware cost, the techniques such as modulation excitation and pulse compression to improve the SNR are not discussed, which means the imaging simulation is assumed to be in an ideal SNR.

The fast broadband beamforming method in [45] is employed to obtain the 3-D image. Figure 4.37 depicts the front view of the 3-D image of using the annular ultralarge ultrasparse UWB 2-D array, which is obtained by projecting the maximum values in each direction. Figure 4.38 depicts the front view of the 3-D image of using the uniformly-spaced ultralarge ultrasparse UWB 2-D array. It can be seen that in Fig. 4.37, the four targets display clearly, including the right angle between A and B, as well as the separation of C and D, while the four targets in Fig. 4.38 cannot be

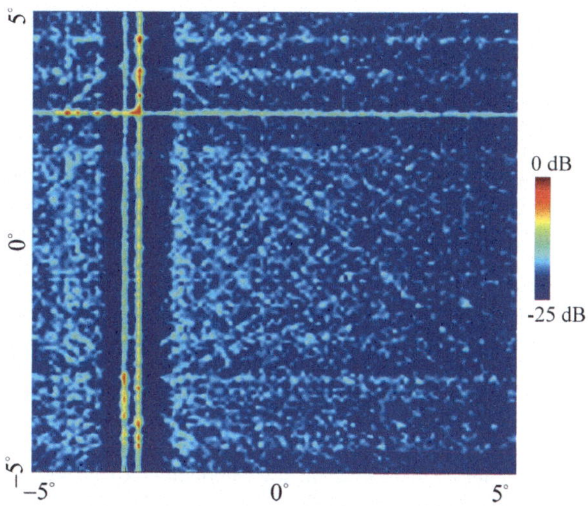

Fig. 4.38 Front view of the 3-D image obtained by using the uniformly-spaced ultralarge ultrasparse UWB 2-D array with the aperture diameter $600\lambda_0$ and 11×11 elements

displayed clearly. Besides, the sidelobe levels of the image in Fig. 4.37 are calculated to be around −20 dB, which means the relatively high quality of the image. To sum up, the imaging results demonstrate that the designed annular ultralarge ultrasparse UWB 2-D array with $B = 70\%$ has a promising performance to achieve a relatively high imaging quality of complicated targets with an angular resolution around 0.1°.

References

1. V. Murino, A. Trucco, Three-dimensional image generation and processing in underwater acoustic vision. Proc. IEEE **88**(12), 1903–1948 (2000)
2. A. Trucco, Thinning and weighting of large planar arrays by simulated annealing. IEEE Trans. Ultrason. Ferroelectr. Freq. Control **46**(2), 347–355 (1999)
3. P. Chen, Y.Y. Zheng, W. Zhu, Optimized simulated annealing algorithms for thinning and weighting large planar arrays in both far-field and near-field. IEEE J. Ocean. Eng. **36**(4), 658–664 (2011)
4. W. Keizer, Large planar array thinning using iterative FFT techniques. IEEE Trans. Antennas Propag. **57**(10), 3359–3362 (2009)
5. P. Chen, B. Shen, L. Zhou, Y. Chen, Optimized simulated annealing algorithm for thinning and weighting large planar arrays. J. Zhejiang Univ-Sci. C (Comput. Electron.) **11**(4), 261–269 (2010)
6. B. Diarra, M. Robini, P. Tortoli, C. Cachard, H. Liebgott, Design of optimal 2-D non-grid sparse arrays for medical ultrasound. IEEE Trans. Biomed. Eng. **60**(11), 3093–3102 (2013)
7. R.L. Haupt, Thinned arrays using genetic algorithms. IEEE Trans. Antennas Propag. **42**(7), 993–999 (1994)
8. K. Chen, X. Yun, Z. He, C. Han, Synthesis of sparse planar arrays using modified real genetic algorithm. IEEE Trans. Antennas Propag. **55**(4), 1067–1073 (2007)
9. D.W. Boeringer, D.H. Werner, Particle swarm optimization versus genetic algorithms for phased arrays systhesis. IEEE Trans. Antennas Propag. **52**(3), 771–779 (2004)

10. A. El-makadema, L. Rashid, A.K. Brown, Geometry design optimization of large-scale broadband antenna systems. IEEE Trans. Antennas Propag. **62**(4), 1673–1680 (2014)
11. J. Córcoles, M.A. González, Efficient combined array thinning and weighting for pattern synthesis with a nested optimization scheme. IEEE Trans. Antennas Propag. **60**(11), 5107–5117 (2012)
12. J. Huo, P. Yang, K. Shi, W. Zhang, Optimal design of random sparse 2-D arrays based on genetic algorithm. ACTA ACUSTICA. **31**(2), 187–191 (2006)
13. C. Chi, Z. Li, Fast computation of wideband beam pattern for designing large-scale 2-D arrays. IEEE Trans. Ultrason. Ferroelectr. Freq. Control **63**(6), 803–816 (2016)
14. J.L. Schwartz, B.D. Steinberg, Ultrasparse, ultrawideband arrays. IEEE Trans. Ultrason. Ferroelectr. Freq. Control **45**(2), 376–393 (1998)
15. V. Murino, A. Trucco, A. Tesei, Beam pattern formulation and analysis for wide-band beamforming systems using sparse arrays. Sig. Process. **56**(2), 177–183 (1997)
16. B.G. Bardsley, D.A. Christensen, Beam patterns from pulsed ultrasonic transducers using linear systems theory. J. Acoust. Soc. Am. **69**(1), 25–30 (1981)
17. G. Cardone, G. Cincotti, P. Gori, M. Pappalardo, Optimization of wideband linear arrays. IEEE Trans. Ultrason. Ferroelectr. Fre. Control. **48**(4), 943–952 (2001)
18. M. Palmese, A. Trucco, An efficient digital CZT beamforming design for near-field 3-D sonar imaging. IEEE J. Ocean. Eng. **35**(3), 584–594 (2010)
19. R.O. Nielsen, *Sonar Signal Process* (Chap. 1–2) (Artech House, Boston, 1991)
20. G. Cardone, G. Cincotti, M. Pappalardo, Design of wide-band arrays for low side-lobe level beam patterns by simulating annealing. IEEE Trans. Ultrason. Ferroelectr. Freq. Control **49**(8), 1050–1059 (2002)
21. J.A. Jensen, N.B. Svendsen, Calculation of pressure fields from arbitrarily shaped, apodized, and excited ultrasound transducers. IEEE Trans. Ultrason. Ferroelectr. Freq. Control **39**(2), 262–267 (1992)
22. J.A. Jensen, Users' guide for the FIELD II program. Release 3.24 (2014)
23. W.J. Hendricks, The totally random versus the bin approach for random arrays. IEEE Trans. Antennas Propag. **39**(12), 1757–1762 (1991)
24. R.E. Davidsen, J.A. Jensen, S.W. Smith, Two-dimensional random arrays for real-time volumetric imaging. Ultrason. Imaging **16**, 143–163 (1994)
25. R.E. Davidsen, S.W. Smith, A multiplexed two-dimensional array for real time volumetric and B-mode imaging, in *Proceedings* of IEEE *Ultrasonics Symposium* (1996), pp. 1523–1526
26. O. Martínez-Graullera, C.J. Martín, G. Godoy, L.G. Ullate, 2D array design based on Fermat spiral for ultrasound imaging. Ultrasonics **50**, 280–299 (2010)
27. A. Trucco, S. Curletto, Flattening the side-lobes of wideband beam patterns. IEEE J. Ocean. Eng. **28**(4), 760–762 (2003)
28. S. Curletto, A. Trucco, On the shaping of the main lobe on wide-band arrays. IEEE Trans. Ultrason. Ferroelectr. Freq. Control **52**(4), 619–629 (2005)
29. T.G. Spence, D.H. Werner, Design of broadband planar arrays based on the optimization of aperiodic tilings. IEEE Trans. Antennas Propag. **56**(1), 76–86 (2008)
30. A. Trucco, M. Palmese, S. Repetto, Devising an affordable sonar system for underwater 3-D vision. IEEE Trans. Instrum. Meas. **57**(10), 2348–2354 (2008)
31. L.Y. Astanin, A.A. Kostylev, Ultrawideband radar measurements analysis and processing. The Institution of Electrical Engineers (1999), pp. 3–5
32. V. Sipal, D. Edwards, B. Allen, Bandwidth requirement for suppression of grating lobes in ultrawideband antenna arrays, in *Proceedings of IEEE International Conference on Ultra-Wideband* (2012), pp. 236–240
33. C. Chi, Z. Li, Q. Li, Ultrawideband underwater real-time 3-D acoustical imaging with ultrasparse arrays. IEEE J. Ocean. Eng. **42**(1), 97–108 (2017)
34. C. Chi, Z. Li, Q. Li, High-resolution real-time underwater 3-D acoustical imaging through designing ultralarge ultrasparse ultra-wideband 2-D arrays. IEEE Trans. Instrum. Meas. **66**(10), 2647–2657 (2017)
35. R.J. Urick, *Principles of Underwater Sound*, 3rd edn. (McGraw-Hill, New York, 1983)

36. B.D. Steinberg, H.M. Subbaram, *Microwave Imaging Techniques* (Wiley, New York, 1991)
37. L.G. Ullate, G. Godoy, O. Martínez, T. Sánchez, Beam steering with segmented annular arrays. IEEE Trans. Ultrason. Ferroelectr. Freq. Control **53**(10), 1944–1954 (2006)
38. W.H. Thorp, Analytic description of the low frequency attenuation coefficient. J. Acoust. Soc. Am. **42**(1), 270 (1967)
39. F.H. Fisher, V.P. Simmons, Sound absorption in sea water. J. Acoust. Soc. Am. **62**(3), 558–564 (1977)
40. T. Misaridis, J.A. Jesen, Use of modulated excitation signals in medical ultrasound. Part I: basic concepts and expected benefits. IEEE Trans. Ultrason. Ferroelectr. Freq. Control. **52**(2), 177–191 (2005)
41. M.A. Richards, *Fundamentals of Radar Signal Processing* (Chap. 4) (McGraw-Hill Companies. Inc., 2005)
42. T. Misaridis, J.A. Jesen, Use of modulated excitation signals in medical ultrasound. Part II: design and performance for medical imaging applications. IEEE Trans. Ultrason. Ferroelectr. Freq. Control **52**(2), 192–207 (2005)
43. B. Haider, P.A. Lewin, K.E. Thomenius, Pulse elongation and deconvolution filtering for medical ultrasonic imaging. IEEE Trans. Ulrason. Ferroelect. Freq. Control **45**(1), 98–113 (1998)
44. J.A. Jensen, J. Mathorne, T. Gravesen, B. Stage, Deconvolution of in-vivo ultrasound b-mode imaging. Ultrason. Imaging **15**, 122–133 (1993)
45. F. Gran, J.A. Jensen, Designing waveforms for temporal encoding using a frequency sampling method. IEEE Trans. Ulrason. Ferroelect. Freq. Contr. **54**(10), 2070–2080 (2007)
46. R.Y. Chiao, X. Hao, Coded excitation for diagnostic ultrasound: a system developer's perspective, in *Proceedings of IEEE Ultrasonics* Symposium (1998), pp. 1639–1644
47. C. Chi, Z. Li, Q. Li, Fast broadband beamforming using nonuniform fast Fourier transform for underwater real-time 3-D acoustical imaging. IEEE J. Ocean. Eng. **41**(2), 249–261 (2016)
48. M. Palmese, A. Trucco, Acoustic imaging of underwater embedded objects: signal simulation for three-dimensional sonar instrumentation. IEEE Trans. Instrum. Meas. **55**(4), 1339–1347 (2006)
49. Y. Han, X. Tian, F. Zhou, R. Jiang, Y. Chen, A real-time 3-D underwater acoustical imaging system. IEEE J. Ocean. Eng. **39**(4), 620–629 (2014)

Chapter 5
Simulation Technique

Abstract The simulation technique for underwater real-time three-dimensional (3-D) acoustical imaging is introduced. This chapter shows how to model the surfaces of the underwater 3-D objects and generate the raw signals acquired by an underwater real-time 3-D acoustical imaging system.

Keywords Acoustic wave propagation · Fast fourier transform · Modeling 3-D objects · Received signals

5.1 Concept and Theory

Developing a practical underwater real-time 3-D acoustical system is a challenging job, which involves two-dimensional (2-D) array design, 3-D imaging algorithms, hardware implementation and other works. To test the performances of imaging algorithms and array design, it is necessary to conduct simulations. Simulations can also help beginners learn the underwater 3-D acoustical imaging technique fast. This chapter discusses how to simulate raw received signals of a real-time 3-D acoustical imaging system.

Generally, typical solid underwater targets have an acoustic impedance which is very different from that of water. When the size of the target is much larger than the wavelength, most of the acoustic energy impinging on the target are reflected and scattered outside. Thus, for underwater real-time 3-D acoustical imaging, only the external surface of a target needs to be considered. Let us consider an underwater 3-D target shown in Fig. 5.1. The surface of the target can be modeled as a dense random grid of discrete scatterers. Rayleigh's scattering law is considered for modeling underwater 3-D target [1–5]. The simulated received signal is obtained by overlapping the responses of the scatterers [1, 2].

Assume that the surface of the target is composed of Q point scatterers. The ith scatterer is at the position $\mathbf{r}_i = (x_i, y_i, z_i)$. The distance of the ith scatterer from the coordinate origin is $r_i = |\mathbf{r}_i|$. Let an acoustic pulse $p(t)$ is emitted by an omnidirectional projector placed in the coordinate origin. The Fourier transform of $p(t)$ is $P(f)$. We assume that a spherical propagation occurs inside a linear isotropic

C. Chi, *Underwater Real-Time 3D Acoustical Imaging*, Signals and Communication Technology, https://doi.org/10.1007/978-981-13-3744-4_5

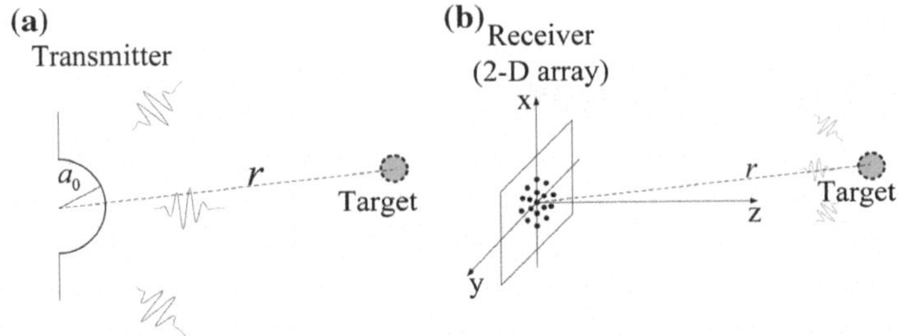

Fig. 5.1 A scene of underwater real-time 3-D acoustical imaging, including **a** an omnidirectional transmitter and **b** a receiver of 2-D array

absorbing medium. The Fourier transform of the pressure field incident on the *i*th scatterer can be computed by the free-space Green's function, expressed as

$$S_{in}(f, \mathbf{r}_i) = \frac{\exp(-\alpha(f)r_i)}{r_i} P(f) \exp\left(\frac{-j2\pi f r_i}{c}\right), \tag{5.1}$$

where $\alpha(f)$ is the absorption coefficient of the medium, f is the frequency, and c is the sound speed. For $\alpha(f)$, the popular equation [6, 7] is

$$\alpha(f) = \frac{0.1 f^2}{(1+f^2)} + \frac{40 f^2}{(4100+f^2)} + 2.75 \times 10^{-4} f^2 + 0.003, \tag{5.2}$$

where $\alpha(f)$ is in dB per kiloyard and f is the frequency in kHz. The distance unit in this book is meter. One yard is equal to 0.914 m. The absorption in dB per meter is expressed as

$$\alpha'(f) = \frac{\alpha(f)}{914}. \tag{5.3}$$

The more accurate and complex model is proposed by Francois and Garrison [8, 9]. Readers who are interested in the Francois-Garrison model, can consider this for obtain a better simulation result. For simplicity, (5.3) is usually employed for simulations of underwater real-time 3-D acoustical imaging.

When we consider a receiver of 2-D array shown in Fig. 5.1(b) at the plane $z = 0$, centered on the coordinate origin, the Fourier transform of the pressure measured at the position $\mathbf{p}_m = (x_m, y_m, 0)$ from the *i*th scatterer can be approximately written as [2]

$$S(f, \mathbf{p}_m)_i \cong \tilde{B}_i \frac{(2\pi f)^2 \exp[-2\alpha(f)r_i]}{c r_i^2} P(f) \exp\left[\frac{-j2\pi f(|\mathbf{p}_m - \mathbf{r}_i| + r_i)}{c}\right], \tag{5.4}$$

where $\tilde{B}_i$ is the parameter describing Rayleigh's scattering process for an unfixed rigid scatterer [2, 3], the dimensions of which are much smaller than the wavelength, expressed as

$$\tilde{B}_i \cong \frac{\alpha_i^3}{3}\left(\frac{\kappa - \kappa_i}{\kappa} + \frac{3\rho_i - 3\rho}{2\rho_i + \rho}\right), \tag{5.5}$$

where α_i is the radius of the scatterer, ρ is the density of the propagation medium, ρ_i is the density of the ith scatterer, κ is the compressibility of the propagating medium, and κ_i is the compressibility of the ith scatterer. Generally, it is assumed that the cosine of the angle between $\mathbf{p} - \mathbf{r}_i$ and $\mathbf{r}_i$ is close to 1. Because Q point scatterers cover the target surface, the Fourier transform of the total pressure at the position $\mathbf{r}$ is the sum of the pressures from all the scatterers, expressed as

$$S(f, \mathbf{p}_m) = \sum_{i=1}^{Q} S(f, \mathbf{p}_m)_i. \tag{5.6}$$

As stated in [2], (5.4) can be used to compute the frequency response of the scene with the sensor at the position $\mathbf{r}$, by imposing $P(f) = 1$.

5.2 Implementation

Computers can only operate in the discrete frequency domain. To calculate the model of (5.4) in computer, the discrete Fourier transform (DFT) needs to be used to obtain the DFT coefficients $P(f_k)$ of $p(t)$. Let f_s be the frequency sampling. The discrete frequency-domain expression of (5.4) can be expressed as

$$S(f_k, \mathbf{p}_m)_i \cong \tilde{B}_i \frac{(2\pi f_k)^2 \exp[-2\alpha(f_k) r_i]}{c r_i^2} P(f_k) \exp\left[\frac{-j2\pi f_k(|\mathbf{p}_m - \mathbf{r}_i| + r_i)}{c}\right], \tag{5.7}$$

where f_k is the kth discrete frequency bin, given by

$$f_k = \frac{k f_s}{K}, \tag{5.8}$$

where K is the length of the DFT. Correspondingly, (5.6) can be rewritten as

$$S(f_k, \mathbf{p}_m) = \sum_{i=1}^{Q} S(f_k, \mathbf{p}_m)_i, \tag{5.9}$$

To simulate received signals sampled by an underwater 3-D acoustical imaging system, the inverse fast Fourier transform (IFFT) is needed. The simulated time-domain sampling signals can be expressed as

$$s(t, \mathbf{r}) = \text{IFFT}[S(f_k, \mathbf{r})], \tag{5.10}$$

where 'IFFT[]' represents the IFFT operation [10].

The distance between the transmitter of an underwater 3-D acoustical imaging system and the target can be over 100 m. If we simulate the received signals in the range from 0 m to 100 m, some computational obstacles will be encountered. An example is given in the following to explain this. We assume that the sound speed c is 1500 m/s. Considering a system with the carrier frequency of 300 kHz and the sampling frequency of 1 MHz, the length of the received signal is $(2 \times 100/1500) \times 10^6 \approx 1.3 \times 10^5$. Generally, the number of the scatterers used in (5.6) is high. The computation of (5.9) costs time. For wideband signals, the number of the frequency bins is also high. The computation needs to be done in each frequency bin. Underwater real-time 3-D acoustical imaging systems include hundreds of sensors generally. We needs to simulate hundreds of raw received signals. Thus, if the length of the IFFT in (5.10) covers the whole range from 0 m to 100 m, the memory required the simulation is over high, and the computing time required by the simulation is over long. Thus, to conduct the simulation efficiently, the following measure should be considered.

It should be noted that the delay term in (5.7), expressed as

$$\tau = \frac{|\mathbf{r} - \mathbf{s}_i| + s_i}{c}. \tag{5.11}$$

Before doing simulations, we can calculate the minimum delay τ_{min} and the maximum delay τ_{max}, according to the imaging 3-D scene and (5.8). The duration of the received signal is

$$\tau_D = \tau_{\text{max}} - \tau_{\text{min}}. \tag{5.12}$$

For the simulations, it is sufficient that the length of the FFT is slightly larger than $\tau_D f_s$. When using the IFFT to obtain the time-domain signals, the circular shift in the time domain arises. Thus, when using (5.10) to compute the time-domain signals, zeros should be padded in the discrete frequency domain, to avoid that the samples near τ_{max} are circularly shifted to the positions near τ_{min}.

References

1. V. Murino, A. Trucco, Three-dimensional image generation and processing in underwater acoustic vision. Proc. IEEE **88**(12), 1903–1948 (2000)
2. M. Palmese, A. Trucco, Acoustic imaging of underwater embedded objects: Signal simulation for three-dimensional sonar instrumentation. IEEE Trans. Instrum. Meas. **55**(4), 1339–1347 (2006)
3. O. George, R. Bahl, Simulation of backscattering of high frequency sound from complex objects and sand sea-bottom. IEEE J. Ocean. Eng. **20**(2), 119–130 (1995)
4. P.C. Pedersen, D.P. Orofino, Modeling of received ultrasound signals from finite planar targets. IEEE Trans. Ultrason. Ferroelectr. Freq. Control **43**(2), 303–311 (1996)
5. P.C. Hines, Theoretical model of acoustic backscatter from a smooth seabed. J. Acoust. Soc. Am. **88**(1), 324–334 (1990)
6. W.H. Thorpe, Description of the low-frequency attenuation coefficient. J. Acoust. Soc. Am. **42**, 270 (1967)
7. R.J. Urick, *Principles of Underwater Sound*, 3rd edn. (McGraw-Hill, New York, 1983)
8. R.E. Francois, G.R. Garrison, Sound absorption based on ocean measurements: Part I: Pure water and magnesium sulfate contributions. J. Acoust. Soc. Am. **72**(3), 896–907 (1982)
9. R.E. Francois, G.R. Garrison, Sound absorption based on ocean measurements: Part II: Boric acid contribution and equation for total absorption. J. Acoust. Soc. Am. **72**(6), 1879–1890 (1982)
10. V.A. Oppenheim, R.W. Schafer, J.R. Buck, *Discrete-Time Signal Processing, Chap. 9* (Prentice-hall, Englewood Cliffs, 1989), pp. 726–745

Chapter 6
System Design

Abstract Developing an underwater real-time three-dimensional (3-D) acoustical imaging system is complex. Only knowing the key techniques: fast beamforming (Chap. 3) and two-dimensional (2-D) array design (Chap. 4) is not enough. The structure of underwater real-time 3-D acoustical imaging systems is introduced in this chapter. The steps of designing an underwater real-time 3-D system are provided.

Keywords Hardware implementation · System structure · System design

6.1 System Structure

Figure 6.1 shows the system structure of underwater real-time 3-D acoustical imaging. The system consists of two sub-systems: transmitting and receiving. The transmitting system includes a digital signal generator, digital-to-analog converter (D/A), power amplifier, transform and transducer. The digital signal generator can be generally realized by storing required waveforms such as continuous narrowband, linear frequency-modulated wideband.

For underwater real-time 3-D acoustical imaging, the receiving 2-D array consists of hundreds of sensors usually, which means the receiving system includes hundreds of sampling channels. In each channel, low-noise amplifier and analog-to-digital converter (A/D) are needed. It should be noted that the amplifiers used in the receiving system should employ a time-gain-control (TGC) circuit. Memory chips are used to store the samples. The fast 3-D imaging algorithms are realized in the digital signal processor. For displaying, 3-D interpolation is also needed. After processing the sampling data, the 3-D acoustic images are outputted to the displayer.

C. Chi, *Underwater Real-Time 3D Acoustical Imaging*, Signals and Communication Technology, https://doi.org/10.1007/978-981-13-3744-4_6

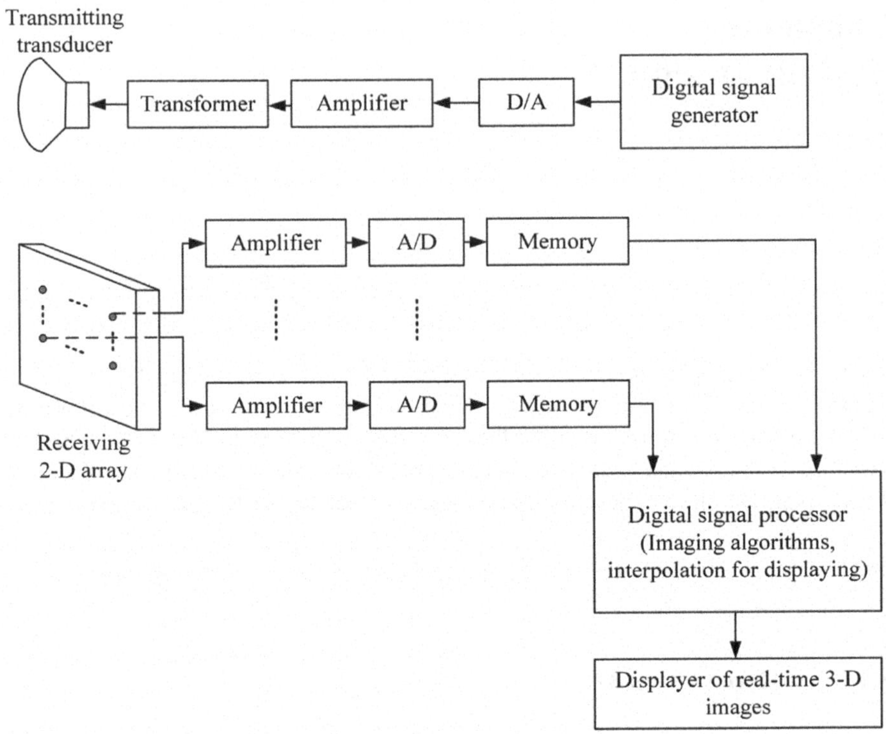

Fig. 6.1 System structure of underwater real-time 3-D acoustical imaging

6.2 Steps of System Design

Before implementing an underwater real-time 3-D acoustical imaging system by hardware and software, we should design first. This book summarizes the system design as five steps:

- Figure out the application requirement
- Determine array parameters
- Design a large sparse 2-D array
- Determine 3-D imaging algorithms
- Choose hardware components and processors

Figure 6.2. illustrates these design steps.

For different applications such as underwater constructions, diver detection and others, the requirements are different. We should consider the specific parameters: system size, power consumption, imaging range, angular resolution and range resolution.

After figuring out the application requirement, we should determine array parameters: aperture size, bandwidth and number of sensors. The aperture size depends

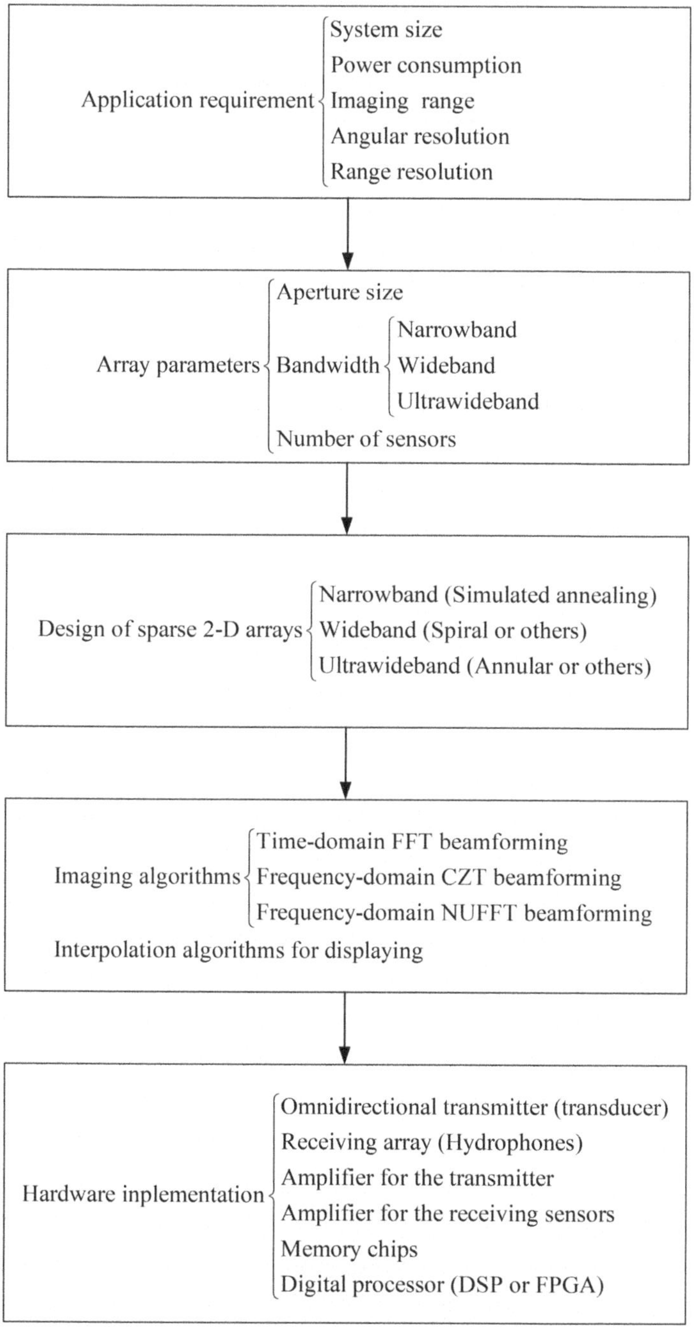

Fig. 6.2 Steps of designing an underwater real-time 3-D acoustical imaging system

on the required angular resolution. Based on the requirement of range resolution, imaging quality and developing cost, we can choose the working bandwidth: narrowband, wideband or ultrawideband (UWB). The number of sensors is set with considering the affordable hardware and signal-to-noise ratio (SNR). The priority should be given to the affordable hardware. The SNR can be compensated by the modulation technique discussed in Chap. 4.

After determining the array parameters, it is necessary to design large sparse 2-D arrays (narrowband, wideband and UWB) with the methods discussed in Chap. 4. Narrowband design can apply the simulated annealing method. The spiral array is a good candidate for the wideband design. The ultrasparse annular array is suitable for the UWB design.

The imaging algorithms (beamforming) should be determined, after finishing designing sparse 2-D arrays. Chapter 3 discusses the beamforming methods for underwater real-time 3-D acoustical imaging. Diverse methods are available for narrowband, wideband and UWB systems.

Finally, we should consider the hardware implementation. Based on the above design parameters, we need to choose a transmitting transducer, D/A, hydrophones for the receiving 2-D array, transmitting amplifier, amplifiers for the receiving sensors, A/D, memory chips and digital processors.

Chapter 7
Existing Challenges and Future Work

Abstract This chapter summarizes the existing challenges for developing next-generation underwater real-time three-dimensional (3-D) acoustical imaging systems. Some interesting research potentials such as compressive sensing, deconvolved conventional beamforming, virtual source technique and others for underwater real-time 3-D acoustical imaging are given.

Keywords Compressive sensing · Deconvolved conventional beamforming · Fast adaptive beamforming · Ultrawideband experiment · Virtual source technique · Wideband 2-D array

7.1 Existing Challenges

As analyzed previously, next-generation underwater real-time 3-D acoustical imaging systems should work in wideband or ultrawideband (UWB). We try to summarize the existing challenges for developing next-generation systems as follows:

(i) It is necessary to develop an experimental wideband underwater 3-D system. Currently, all the practical underwater real-time 3-D systems are narrowband. Narrowband systems only need to process received signals at single frequency bin. The narrowband design methods of large underwater sparse 2-D arrays are relatively mature. The development of a wideband real-time underwater 3-D acoustical imaging system is much more complex than that of a narrowband one. When developing wideband 3-D real-time imaging system, the problems faced will be more demanding.

(ii) We require a higher angular resolution. Angular resolutions of the existing underwater 3-D acoustical systems are around 1°. For the angular resolution of 1°, the lateral resolution is 1.7 m at a distance of 100 m. We can find that it will be difficult to image the small target at large distances. Thus, improving the angular resolution is needed.

(iii) A theoretical criterion of deciding how sparse 2-D acoustical arrays should be is needed. The existing narrowband array design methods mainly consider minimizing the number of elements and optimizing the weights under the directivity

C. Chi, *Underwater Real-Time 3D Acoustical Imaging*, Signals and Communication Technology, https://doi.org/10.1007/978-981-13-3744-4_7

requirements. The existing wideband and UWB design methods are based on the specific geometries with the number of sensors fixed. However, the measure of a good 2-D array is not always: the sparser, the better. There should be different criteria for different situations. To obtain a good 3-D underwater image, a high signal-to-noise ratio (SNR) is needed, which the theoretical criterion should consider.

(iv) Wideband real-time 3-D imaging systems need to reduce the high computational load continuously. From the author's view, for narrowband real-time 3-D imaging systems, the computational load has not been a problem. However, for wideband systems, the computational load is still a problem. Wideband systems must be able to process a high number of discrete frequencies in real time. Fortunately, some fast wideband 3-D beamforming methods such as chirp zeta transform [1] and nonuniform fast Fourier transform [2] have been made available. More efficient beamforming methods are needed in the future.

7.2 Future Research Potentials

For developing next-generation real-time 3-D acoustical imaging systems, we find that there are some interesting research potentials. The future research potentials are summarized as follows:

(i) Introducing the compressive sensing techniques [3–6] into underwater real-time 3-D acoustical imaging. The compressive sensing techniques have been applied to many imaging fields such as medical ultrasound 3-D imaging [7], computed tomography [8], synthetic aperture radar [9] and magnetic resonance imaging [10]. Currently, the compressive sensing techniques have not been considered in underwater real-time 3-D acoustical imaging. Hence, the compressive sensing techniques for underwater 3-D acoustical imaging deserve greater attention.

(ii) Applying the virtual source technique [11–14] to improve the transmitting power for wideband or UWB underwater 3-D systems. Figure 7.1 shows the schematic of the virtual source technique. The focal point of the transducers is taken as a virtual source, which transmits a spherical wave in a limited angular range [14]. Several transducers which work together can help enhance the transmitting power. For wideband or UWB underwater real-time 3-D acoustical imaging systems, transmitting enough power by a single transducer might be a problem. The virtual source technique can be a viable means to increase the transmitting power.

(iii) Employing the deconvolved conventional beamforming techniques [15–19] to improve the angular resolution. The deconvolved conventional beamforming techniques have been applied to estimate the direction of arrival [15, 16], and mapping of acoustic sources [17–19]. Figure 7.2 shows the outputs of the conventional beamforming and deconvolved conventional beamforming. We can

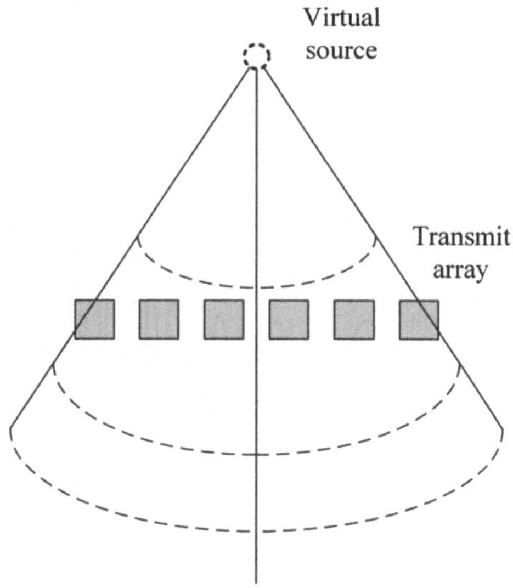

Fig. 7.1 Schematic of the virtual source technique

find that the deconvolved conventional beamforming obtains a much higher angular resolution than the conventional beamforming. It should be noted that the deconvolved conventional beamforming is able to work well in a good SNR. Because underwater real-time 3-D acoustical imaging systems are active, it is not hard to meet the SNR requirement. In addition, we need to consider the computational load for real-time 3-D imaging. Some fast deconvolved conventional beamforming algorithms, such as the DAMAS2 algorithm of [17, 18], are available. With the trend of declining cost and accelerating processing speed, it is convinceable that the deconvolved conventional beamforming techniques are able to be applied to underwater real-time 3-D acoustical imaging in the future.

(iv) Designing better wideband large sparse 2-D arrays. For underwater real-time 3-D acoustical imaging, designing wideband large sparse 2-D arrays is complex and crucial. Some heuristic algorithms [20, 21] have been applied to design narrowband 2-D arrays of underwater 3-D imaging. However, these algorithms have not been applied to design wideband 2-D arrays by using wideband beam pattern. Wideband 2-D arrays with higher sparsity and better directivity are necessary for developing next-generation underwater 3-D imaging systems.

(v) Designing Better UWB 2-D arrays. Reference [22] and Chap. 4 reveal that for underwater real-time 3-D acoustical imaging, UWB ultra-sparse arrays are feasible and can obtain ultra-low hardware cost. An annular 2-D UWB array was employed in [22]. Unfortunately, how to design optimal UWB ultra-sparse 2-D arrays for the application has not been researched. Better UWB 2-D arrays can help achieve better imaging quality and lower hardware cost. Thus, designing better UWB 2-D arrays is a potential direction for the researchers in the future.

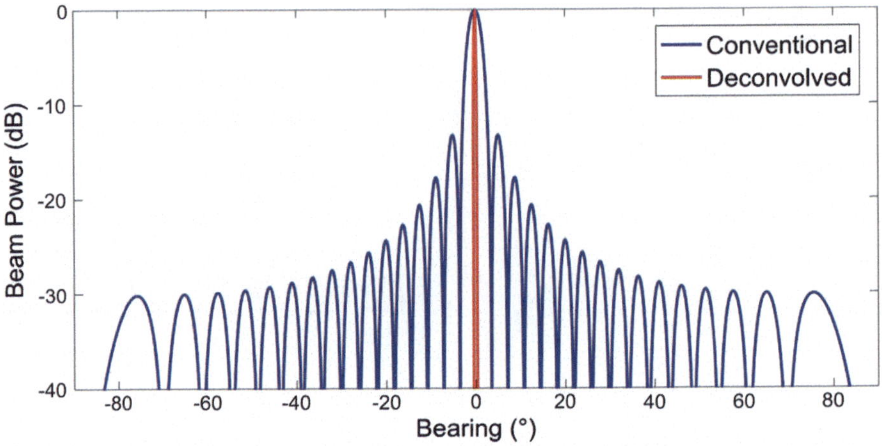

Fig. 7.2 Comparison of the outputs of the conventional beamforming and deconvolved conventional beamforming, for a linear array

(vi) Doing experimental studies for underwater wideband and UWB real-time 3-D acoustical imaging. To date, all reports on underwater wideband and UWB real-time 3-D acoustical imaging are at the simulation stages. None of the experimental studies have been reported. To develop a commercial underwater wideband or UWB 3-D acoustical imaging system, early experimental studies are necessary. More importantly, doing experiments can help us find more problems and more research potentials.

References

1. M. Palmese, A. Trucco, An efficient digital CZT beamforming design for near-field 3-D sonar imaging. IEEE J. Ocean. Eng. **35**(3), 584–594 (2010)
2. C. Chi, Z. Li, Q. Li, Fast broadband beamforming using nonuniform fast Fourier transform for underwater real-time 3-D acoustical imaging. IEEE J. Ocean. Eng. **41**(2), 249–261 (2016)
3. D.L. Donoho, Compressed sensing. IEEE Trans. Inf. Theory **52**(4), 1289–1306 (2006)
4. E.J. Candes, T. Tao, Near-optimal signal recovery from random projections: universal encoding strategies? IEEE Trans. Inf. Theory **52**(12), 5406–5425 (2006)
5. Y.C. Eldar, G. Kutyniok, *Compressed Sensing: Theory and Applications* (Cambridge University Press, Cambridge, 2012)
6. N. Wagner, Y.C. Eldar, Z. Friedman, Compressed beamforming in ultrasound imaging. IEEE Trans. Signal Process. **60**(9), 4643–4657 (2012)
7. A. Burshtein, M. Birk, T. Chernyakova, A. Eilam, A. Kempinski, Y.C. Eldar, Sub-nyquist sampling and Fourier domain beamforming in volumetric ultrasound imaging. IEEE Trans. Ultrason., Ferroelectr., Freq. Control **63**(5), 703–717 (2016)
8. K. Choi, J. Wang, L. Zhu, T. Suh, S. Boyd, L. Xing, Compressed sensing based cone-beam computed tomography reconstruction with a first-order method. Med. Phys. **37**(9), 5113–5125 (2010)

9. V. M. Patel, G. R. Easley, D. M, Healy Jr, R. Chellappa, Compressed synthetic aperture radar. IEEE J. Sel. Top. Signal Process. **4**(2), 244–254 (2010)
10. M. Lustig, D.L. Donoho, J.M. Santos, J.M. Pauly, Compressed sensing MRI. IEEE Signal Process. Mag. **25**(2), 72–82 (2008)
11. C.H. Frazier, W.D. O'Brien, Synthetic aperture techniques with a virtual source element IEEE Trans. Ultrason., Ferroelect., Freq. Control. **45**(1), 196–207 (1998)
12. M.H. Bae, M.K. Jeong, A study of synthetic-aperture imaging with virtual source elements in B-mode ultrasound imaging systems. IEEE Trans. Ultrason., Ferroelect., Freq. Control **47**(6), 1510–1519 (2000)
13. E. Moghimirad, C.A.V. Hoyos, A. Mahloojifar, B.M. Asl, J.A. Jensen, Synthetic aperture ultrasound Fourier beamformation sing virtual sources. IEEE Trans. Ultrason., Ferroelect., Freq. Control **63**(12), 2018–2030 (2016)
14. M. Yu, Y. Li, T. Ma, K.K. Shung, Q. Zhou, Intravascular ultrasoumd imaging with virtual source synthetic aperture (VSSA) focusing and coherence factor weight (CFW). IEEE Trans. Med. Imaging **36**(10), 2171–2178 (2017)
15. T.C. Yang, Deconvolved conventional beamforming for a horizontal line array. IEEE J. Ocean. Eng. **43**(1), 160–172 (2018)
16. T.C. Yang, Performance analysis of superdirectivity of circular arrays and implications for sonar systems. IEEE J. Ocean. Eng. **99**, 1–11 (2018). (Early access)
17. K. Ehrenfried, L. Koop, Comparison of iterative deconvolution algorithms for the mapping of acoustic sources. AIAA J **45**(7), 1584–1595 (2007)
18. R.P. Dougherty, Extensions of DAMAS and benefits and limitations of deconvolution in beamforming, in *AIAA Paper*, 2005–2961 (2005)
19. T.F. Brooks, W.M. Humphreys, A deconvolution approach for the mapping of acoustic sources (DAMAS) determined from phased microphone arrays, in *AIAA Paper*, pp. 2004–2954 (2004)
20. A. Trucco, Thinning and weighting of large planar arrays by simulated annealing. IEEE Trans. Ultrason. Ferroelectr. Freq. Control **46**(2), 347–355 (1999)
21. P. Chen, Y.Y. Zheng, W. Zhu, Optimized simulated annealing algorithms for thinning and weighting large planar arrays in both far-field and near-field. IEEE J. Ocean. Eng. **36**(4), 658–664 (2011)
22. C. Chi, Z. Li, Q. Li, Ultrawideband underwater real-time 3-D acoustical imaging with ultra-sparse arrays. IEEE J. Ocean. Eng. **42**(1), 97–108 (2017)

The manufacturer's authorised representative in the EU is Springer Nature Customer Service Centre GmbH, Europaplatz 3, 69115 Heidelberg, Germany. If you have any concerns regarding our products, please contact ProductSafety@springernature.com

Printed and bound by CPI Group (UK) Ltd, Croydon, CR0 4YY

25/09/2026

02222892-0016